Underwater Acoustic Technologies for Sustainable Fisheries

Underwater Acoustic Technologies for Sustainable Fisheries

Guest Editors

Jianfeng Tong
Yong Tang
Tohru Mukai

MDPI

Basel • Beijing • Wuhan • Barcelona • Belgrade • Novi Sad • Cluj • Manchester

Guest Editors
Jianfeng Tong
College of Marine Living Resource Sciences and Management
Shanghai Ocean University
Shanghai
China

Yong Tang
College of Marine Living Resource Sciences and Management
Shanghai Ocean University
Shanghai
China

Tohru Mukai
Faculty of Fisheries Sciences
Hokkaido University
Sapporo
Japan

Editorial Office
MDPI AG
Grosspeteranlage 5
4052 Basel, Switzerland

This is a reprint of the Special Issue, published open access by the journal *Fishes* (ISSN 2410-3888), freely accessible at: https://www.mdpi.com/journal/fishes/special_issues/17JZYT7418.

For citation purposes, cite each article independently as indicated on the article page online and as indicated below:

Lastname, A.A.; Lastname, B.B. Article Title. *Journal Name* **Year**, *Volume Number*, Page Range.

ISBN 978-3-7258-6029-6 (Hbk)
ISBN 978-3-7258-6030-2 (PDF)
https://doi.org/10.3390/books978-3-7258-6030-2

Contents

fishes

MDPI

Editorial

Underwater Acoustic Technologies for Sustainable Fisheries

Jianfeng Tong [1,*], Yong Tang [1] and Tohru Mukai [2]

[1] College of Marine Living Resource Sciences and Management, Shanghai Ocean University, Shanghai 201306, China; y-tang@shou.edu.cn
[2] Faculty of Fisheries Sciences, Hokkaido University, Hakodate 0418611, Hokkaido, Japan; mukai@fish.hokudai.ac.jp
* Correspondence: jftong@shou.edu.cn

Underwater acoustic technologies have emerged as indispensable tools for advancing research, management, and conservation across aquatic ecosystems. Their capacity to reveal critical information on animal behavior, migration patterns, population dynamics, and habitat use underscores their value not only for sustainable fisheries but also for the protection of threatened and endangered species. In the face of growing pressures from anthropogenic impacts and climate change, such technologies represent essential means of supporting resilient and sustainable aquatic systems. This Special Issue aims to showcase cutting-edge applications of underwater acoustic technologies, from supporting sustainable fisheries to improving acoustic monitoring precision and efficiency and advancing broader ecosystem research, highlighting their growing importance in aquatic science and conservation.

This Special Issue brings together nine significant contributions that collectively highlight the potential of underwater acoustic technologies to advance sustainable fisheries and inform aquatic ecosystem studies. Through case studies and comprehensive reviews, these studies enhance our understanding of aquatic organisms and provide practical tools and methodologies for fisheries' resource assessment and species-level ecological research. By illustrating applications ranging from target strength estimation and broadband acoustic techniques to behavioral analysis, these contributions demonstrate the versatility and growing importance of acoustic approaches in aquatic science and conservation.

Target strength (TS) is a fundamental parameter in fisheries' acoustic assessments, as it directly influences the species identification and the estimation of biomass. Accurate TS measurements are therefore essential for reliable resource assessment and the effective application of acoustic techniques in fisheries research. In this Special Issue, three studies focus specifically on improving TS estimation and understanding the factors that affect acoustic scattering. These contributions collectively address technical challenges in broadband quantification, validate theoretical models such as the Kirchhoff–Ray Mode (KRM) against empirical data, and provide specific TS–body length relationships essential for accurate biomass assessment. Ai et al. (article 1) developed a broadband scattering quantification method for single fish using split-beam echosounders, integrating echo field theory, transducer equivalent circuits, and signal processing to enhance the generalizability across sonar systems and reduce the reliance on calibration standards. Zhu et al. (article 2) examined the TS in Japanese mackerel (*Scomber japonicus*), assessing the effects of freezing on swimbladder morphology and TS and demonstrating strong agreement between ex situ measurements and KRM model simulations, while providing empirical broadband TS–length relationships. Meng et al. (article 3) analyzed the TS in large yellow croaker (*Larmichthys crocea*), revealing its dependence on the tilt angle, frequency, and body length, and establishing least-squares fitted TS–length equations superior to conventional b20 values for accurate acoustic assessment. Collectively, these studies highlight how biological and environmental factors—including swimbladder integrity, tilt angle, and sample preservation—affect acoustic scattering, providing a robust foundation for species-level identification and quantitative fisheries resource evaluation.

Building on the foundational role of TS in quantitative assessments, underwater acoustics also provide powerful tools for investigating fish distributions and monitoring ecological change at broader ecosystem scales. Recognizing the crucial intermediary role of Antarctic silverfish (*Pleuragramma antarcticum*) in the Antarctic food web, Lee et al. (article 4) investigated their spatiotemporal distribution in the Ross Sea, using the KRM backscattering model to estimate the TS and mean volume backscattering strength (MVBS) to reveal vertical and horizontal patterns. Their findings showed that most juveniles concentrated around 100 m depth, near sea ice and polynya waters, while also highlighting the need for refined algorithms to separate silverfish from krill. Li et al. (article 5) explored the temporal and spatial dynamics of fish resources in the confluence of Poyang Lake and the Yangtze River during the early stage of China's ten-year fishing ban. Acoustic surveys indicated higher fish densities in the confluence zone during the high-water period and aggregation in deeper river channels during the low-water period. The study revealed increases in both fish density and body length, providing strong evidence for the policy's effectiveness in stock recovery. They also pointed out the ongoing challenge in accurately estimating the fish total length from the TS, emphasizing the need for incorporating diverse equations that consider the fish shape, swimbladder size, and transducer–fish positional relationships for enhanced accuracy in future research. The review by González-Máynez et al. (article 6) strongly supports the relevance of such studies in fresh and shallow waters. It indicates that inland waters, including lakes, rivers, streams, and reservoirs, play a significant role in sustaining riverine fish abundance and serve as crucial reservoirs of biodiversity. The review further highlights the practical benefits of acoustic methods in these environments, noting that echosounders can be readily installed on small boats, thereby presenting new opportunities for evaluation. Additionally, it underscores the value of acoustic methods for providing non-invasive long-term observations in sensitive regions such as Marine Protected Areas (MPAs). This characteristic makes acoustic approaches particularly applicable for assessing the impact of conservation measures such as fishing bans.

The studies discussed above primarily employed scientific echosounders. In addition, acoustic cameras represent another important technological approach in fishery acoustics. Wang et al. (article 7) assessed the use of an acoustic camera named Adaptive Resolution Imaging Sonar (ARIS) for monitoring large jellyfish in Liaodong Bay. Their results demonstrated the ability of acoustic imaging to identify species based on the size, shape, and movement, while yielding abundance estimates substantially higher than those from net sampling. The method also proved advantageous in shallow, turbid, or nighttime conditions, despite some limitations related to blind zones and the minimum detectable size.

Extending the application of acoustic camera technology, Shen et al. (article 8) combined artificial intelligence with another type of acoustic camera, the Dual-frequency Identification Sonar (DIDSON), to develop a fish target identification and counting method. This approach, utilizing the YOLOv5 deep learning model combined with the DeepSort tracking algorithm, achieved high identification accuracies and significantly reduced the processing time compared to manual or traditional software analysis. By minimizing human effort and bias, this study highlights the potential of combining acoustic imaging with AI to enhance the precision, efficiency, and objectivity of aquatic ecosystem monitoring and management.

Passive acoustic monitoring is another acoustic technology that significantly contributes to aquatic ecosystem research, expanding beyond active fishery acoustics. This approach, which involves listening to the sounds produced by aquatic animals, provides invaluable insights into their presence, behavior, and the health of their environment without direct physical intervention. Chen et al. (article 9) provide a compelling example of passive acoustic monitoring through their study on the vocalization patterns and echolocation signals of the endangered Yangtze Finless Porpoise in captivity. They showed that the signal characteristics vary with behavioral states, enabling precise target discrimination. These insights are valuable for conservation, guiding strategies to reduce human disturbance and assess interactions with fishing gear, and demonstrate the broader potential of passive acoustics for monitoring endangered species and supporting ecosystem management.

Collectively, the studies presented in this Special Issue highlight the pivotal role of underwater acoustic technologies in supporting sustainable fisheries. From precise biomass estimation using echosounders to species identification and behavioral monitoring via acoustic cameras and non-invasive assessment of endangered species through passive acoustic monitoring, these contributions demonstrate how acoustic approaches can enhance both resource management and ecological understanding. Looking ahead, continued advances in acoustic instrumentation, modeling, and artificial intelligence will further strengthen the capacity for the effective long-term monitoring of fish populations and aquatic ecosystems. By improving species discrimination and enabling automated data processing, these technological developments pave the way for broader applications. Coupling underwater acoustics with emerging observation platforms—including aquaculture net–cage monitoring systems, autonomous aerial, surface, and underwater vehicles, and moored or drifting buoys—will enable more comprehensive and adaptive monitoring of aquatic ecosystems. Such developments have the potential to advance ecosystem-based management and reinforce the role of acoustics in achieving sustainable fisheries under changing environmental conditions.

Conflicts of Interest: The authors declare no conflicts of interest.

List of Contributions:

1. Ai, Q.; Li, H.; Yao, J.; Li, C.; Tao, J. Broadband Scattering Characteristic Quantization Technique for Single Fish Based on a Split-Beam Echosounder. *Fishes* **2024**, *9*, 12. https://doi.org/10.3390/fishes9010012.
2. Zhu, Z.; Tong, J.; Xue, M.; Qiu, C.; Lyu, S.; Liu, B. Investigations on Target Strength Estimation Methods: A Case Study of Chub Mackerel (*Scomber japonicus*) in the Northwest Pacific Ocean. *Fishes* **2024**, *9*, 307. https://doi.org/10.3390/fishes9080307.
3. Meng, J.; Tang, Y.; Sun, L.; Lin, L.; Li, Y.; Miao, X.; Liu, S.; Song, P. Acoustic Target Strength Measurement of *Larmichthys crocea* Based on the Kirchhoff-Ray Mode Model. *Fishes* **2024**, *9*, 424. https://doi.org/10.3390/fishes9110424.
4. Lee, S.; Oh, W.; La, H.S.; Son, W.; Kim, J.-H.; Lee, K. Spatiotemporal Distribution of Antarctic Silverfish in the Ross Sea, Antarctica. *Fishes* **2024**, *9*, 47. https://doi.org/10.3390/fishes9020047.
5. Li, H.; Yu, X.; Wu, B.; Yu, L.; Wang, D.; Wang, K.; Wang, S.; Chen, D.; Li, Y.; Duan, X.; et al. Temporal and Spatial Distribution Characteristics of Fish Resources in a Typical River–Lake Confluence Ecosystem During the Initial Period of Fishing Ban. *Fishes* **2024**, *9*, 492. https://doi.org/10.3390/fishes9120492.
6. González-Máynez, V.E.; Morales-Bojórquez, E.; Nevárez-Martínez, M.O.; Villalobos, H. Application of Fisheries Acoustics: A Review of the Current State in Mexico and Future Perspectives. *Fishes* **2024**, *9*, 387. https://doi.org/10.3390/fishes9100387.
7. Wang, B.; Liu, X.; Dong, J.; Wang, A.; Feng, C.; Xu, Y.; Zhang, D.; Zhao, Z. Evaluating the Effectiveness of an Acoustic Camera for Monitoring Three Large Jellyfish Species in the Coastal Waters of Liaodong Bay, China. *Fishes* **2025**, *10*, 105. https://doi.org/10.3390/fishes10030105.

8. Shen, W.; Liu, M.; Lu, Q.; Yin, Z.; Zhang, J. A Fish Target Identification and Counting Method Based on DIDSON Sonar and YOLOv5 Model. *Fishes* **2024**, *9*, 346. https://doi.org/10.3390/fishes9090346.
9. Chen, J.; Liang, H.; Lin, D.; Zhang, J.; Li, D.; Ye, K.; Lu, W.; Liu, K. Vocalization Pattern and Echolocation Signal Characteristics of Yangtze Finless Porpoise (*Neophocaena asiaeorientalis asiaeorientalis*) in Captivity. *Fishes* **2024**, *9*, 119. https://doi.org/10.3390/fishes9040119.

fishes

MDPI

Article

Broadband Scattering Characteristic Quantization Technique for Single Fish Based on a Split-Beam Echosounder

Qiuming Ai [1,2], Haisen Li [1,3,4,*], Jin Yao [1], Chao Li [5] and Jiangping Tao [1]

[1] College of Underwater Acoustic Engineering, Harbin Engineering University, Harbin 150001, China; kiana@hrbeu.edu.cn (Q.A.); yaojin1982@hrbeu.edu.cn (J.Y.); jptao@mail.ihe.ac.cn (J.T.)
[2] Nanhai Institute, Harbin Engineering University, Sanya 572024, China
[3] National Key Laboratory of Underwater Acoustic Technology, Harbin Engineering University, Harbin 150001, China
[4] Key Laboratory of Marine Information Acquisition and Security, Harbin Engineering University, Ministry of Industry and Information Technology, Harbin 150001, China
[5] SANYA Offshore Oil & Gas Research Institute, Northeast Petroleum University, Sanya 572024, China; chaoli@hrbeu.edu.cn
* Correspondence: hsenli@126.com

Abstract: The utilization of broadband quantization data increases the possibility of practical applications for fish target recognition based on the acoustic scattering theory. However, the quantification of broadband data is more complex than that of narrowband systems, requiring consideration of the broadband characteristics of split-beam transducers, seawater absorption, and circuits. This paper elucidates the scatterer acoustic field equation, transducer power equation, and signal processing flow in split-beam broadband quantization technology for engineering applications. A broadband calculation model based on transducer parameters is proposed to enhance the generalization ability of broadband quantization technology to different types of sonar. The classical echo integration method is combined with Fourier transform to meet the requirements of target strength (TS) estimation under broadband signals. This paper includes a series of experiments to prove the rationality and effectiveness of the method. The results demonstrate that the provided calculation model can more accurately reflect the backscattering characteristics of the scatterer, but certain errors remain. This article analyzes the sources of errors and validates the effectiveness of the new TS calculation method.

Keywords: split-beam echosounder; broadband technology; target strength; Helmholtz-Thevenin circuit; echo integration; Fourier transform; fisheries acoustics

Key Contribution: By combining scatterer acoustic field theory, an equivalent circuit for the transducer, and digital signal processing technology, the broadband TS calculation principle for fish is derived in detail, and a new broadband TS estimation method is proposed.

1. Introduction

Acoustic detection technology based on scientific echosounders is widely used in the field of fishery resource investigation [1–9]. This type of echosounder performs vertical or horizontal detection underwater by transmitting and receiving acoustic waves [1]. The sound pressure is converted to an electrical signal by the transducer and amplified by the circuits, and echo integration is used to analyze the target scattering characteristics [2–8]. The split-beam method is an improvement based on single-beam technology [9] and it is one of the mainstream methods for measuring the backscattering cross-section of marine organisms [5,7]. The principle is that by splitting the receiving plane array into four quadrants [2,3,5–7], and then employing the principle of time delay positioning [10], Assuming that the values of the beam pattern at all angles in space are known, the influence of the beam pattern can be removed based on the calculated target position [2,5–7]. In principle, the TS of the detected object can be calculated through a single echo calculation. Simrad EK500 and EK60 are narrow-band split-beam echosounders commonly used in fishery acoustic surveys [5,11–13]. Among these, the EK500 operation manual proposes TS calculation based on the power budget equation. The formula is deformed based on the radar equations, and the TS value is derived, based on the spherical scattering cross section [14]. The spherical scattering cross section corresponds to the radar scattering cross section in the radar equation [15,16], which needs to be divided by the spatial angle to convert it into the acoustic backscattering cross section [17]. Lunde derived the calculation formulas of narrow-band devices such as EK500 and EK 60 in detail by using acoustic models [11]. Although the formula is reasonable, the definition of transducer array gain G is too complex. Most calculation methods rely on equipment and calibration parameters [1,3,5,6,11–13] and do not mention the processing of the beam pattern. EK80 is currently one of the most mainstream broadband split-beam echo sounders and is considered to be the next-generation device to replace EK60 [12,13]. Most broadband split-beam research is based on this assumption. Lars Nonboe Andersen provided a comprehensive introduction to the quantitative processing of broadband data using the EK80 broadband system [7]. The TS calculation formula was derived based on Lunde's formula, which lacks an analysis of the processing of the beam pattern under the broadband, and the transducer gain parameter G in the equation relies on the broadband calibration of the standard scattering sphere [7,12,13,18]. The calibration of broadband equipment is more complicated. Due to the multiple resonances of the sphere within the broadband [19], a rigid sphere of specific material is generally selected as a reference, ensuring the effectiveness of the calibration of the broadband equipment [20]. Dezhang Chu also stated that during the process of designing calibration experiments for broadband systems, calibration is essential in order to obtain high-quality acoustic data [21]. Broadband systems bring more information to fishery acoustics [7,12,22], but data processing is complex and unclear. Although most of the formulas have been proven to be reasonable [7,11], the current calculation methods for broadband systems do not provide a complete acoustic scheme. An over-reliance on calibration information makes it impossible to obtain a valid absolute estimate of the TS value in some scenarios in which there are no standard scatterers. To combine broadband technology with split-beam echo sounders, this article takes the scientific echosounder EK80 as the research object and carries out the following research.

The main purpose of this article is to:

1. Derive the radiation and single target scattering model of the circular piston transducer under far-field conditions. The calculation formula for the backscattering cross section under a single target is given by combining the Helmholtz-Thevenin equivalent circuits model and the acoustic parameters of the transducer.
2. Establishes a processing flow for broadband data under a single target and derives the echo integration under broadband based on Fourier transform.
3. Using EK80 as the experimental object, this research designs several transducer parameter test experiments and a copper ball TS measurement experiment to compare and verify the accuracy of the calculation model.

2. Materials and Methods

2.1. Single-Target Scattering Model and Backscattering Cross Section in the Far Field

The derivation is based on the free field. Both the scatterer and the transducer satisfy far-field conditions, and the finite amplitude effect is ignored [23,24]. The transducer is a transceiver and satisfies reciprocity.

Figure 1 shows a simplified detection scenario in a typical monostatic configuration, where the transducer is a circular piston transducer, and the spherical coordinate system is established based on the equivalent acoustic center of the transducer plane. Considering absorption loss, the sound pressure p of the reference point (r, θ) in the sound field is shown in Equation (1) [25].

$$p(r,\theta) = j\rho_w c_w k u_0 a^2 e^{-jkr} \frac{10^{-\alpha r/20}}{r} \frac{J_1(ka\sin(\theta))}{ka\sin(\theta)} \tag{1}$$

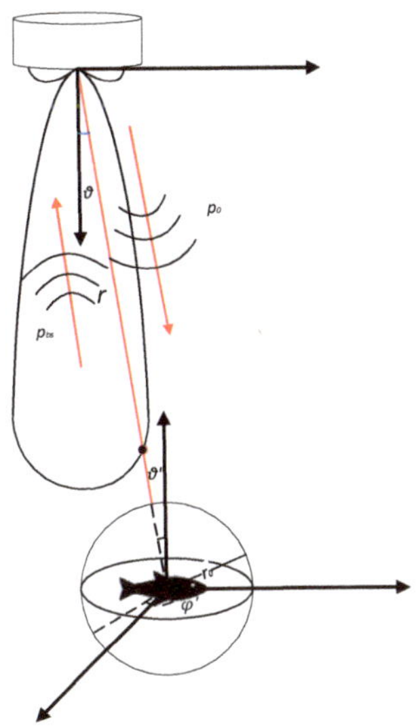

Figure 1. Schematic of a single fish backscatter acoustic system, with an electroacoustic transducer as the transmitter and receiver of the acoustic signal.

Among these, ρ_w is the density of water, c_w is the sound velocity of water, k is the wave number, u_0 is the particle vibration velocity, a is the diameter of the transducer, α Is the absorption coefficient of the medium, and J_1 is the first-order Bessel function. For the convenience of subsequent analysis, we take the reference distance r_0 and set the sound pressure amplitude as p_0, ignoring the time-harmonic factor e^{-jkr}, to obtain the radiation sound pressure amplitude of the transducer in the sound field as shown in Equation (2) [24], Equation (3) is the expression of the beam pattern [25].

$$|p(r,\theta)| = p_0 \frac{10^{-\alpha r/20}}{r} b(\theta) \tag{2}$$

$$b(\theta) = \frac{J_1(ka\,\sin(\theta))}{ka\,\sin(\theta)} \tag{3}$$

It is assumed that the object in Figure 1 is a single object of any shape or material, or a collection of various types of objects of different types of materials. The overall volume is limited to meet the requirements that it should be smaller than the first Fresnel zone of the sound beam, which can be completely engulfed by the sound beam. Objects that meet these requirements can be considered as a single target for analysis, and their scattering characteristics can be regarded as scattering from point-like objects [17]. It is assumed that the far field conditions for point scattering are met. Taking the backscattered sound pressure amplitude p_{bs0} at the reference distance r_0 as the initial value before the spherical diffusion of the sound pressure, the scattered wave is received by the transducer in the form of a plane wave. Then the sound pressure p_{bs} generated by the scatterer at the sound center of the transducer can be obtained, as shown in Equation (4).

$$p_{bs} = \frac{p_{bs0}}{r} 10^{-\alpha r/20} \quad (4)$$

Based on the physical definition of the backscattering cross-section mentioned previously [26], Equation (5) can be obtained.

$$\frac{p_{bs0}}{|p(r,\theta)|} = \frac{\sqrt{\sigma_{bs}}}{r_0} \quad (5)$$

According to Equations (2)–(5), the final amplitude of sound pressure generated on the surface of the transducer can be obtained using Equation (6).

$$p_{bs} = p_{bs0}\frac{10^{-\alpha r/20}}{r} = p_0 \frac{10^{-\alpha r/10}}{r^2} b(\theta)\sqrt{\sigma_{bs}} \quad (6)$$

To calculate the backscattering cross-section of the target, Equation (6) is deformed, and the backscattering cross-section with respect to the square ratio of the sound pressure amplitude is obtained, as shown in Equation (7).

$$\sigma_{bs} = \left(\frac{p_{bs}}{p_0}\right)^2 \frac{r^4 10^{-2\alpha r/10}}{b^2(\theta)} \quad (7)$$

Equation (7) is then converted into TS [17,26], as shown in Equation (8).

$$TS = 20\log(p_{bs}) - 20\log(p_0) + 40\log(r) + 2\alpha r - 20\log(b(\theta)) \quad (8)$$

Theoretically, by measuring the emission sound pressure value and the scattering sound pressure value, calculating the spatial coordinates of the scattering target, and finally combining them with absorption loss of the medium, the backscattering cross section can be acquired, and then the TS can be obtained. In the next section, the connection between the sound pressure amplitude and the circuit system will be introduced and the theoretical calculation equation of the TS in broadband scenarios will be provided.

All the symbols and variables in Section 2.1 are presented in Table A1 in Appendix A, containing units and a brief description.

2.2. *Equivalent Model of Helmholtz-Thevenin Circuit for Transducers*

In order to obtain the relationship between the ratio of sound pressure to the circuit, we must focus on describing the conversion performance of the acoustic and electrical signals of the transducer, which is defined as the sensitivity of the transducer. It is assumed that the transducer satisfies reciprocity, is passive, that the voltage it transmits and receives is linear in relation to the sound pressure, and that the free-field condition is satisfied. The transmitting sensitivity s_u and receiving sensitivity m_u of the transducer are defined by voltage as follows [27,28]:

$$s_u = \frac{p_0}{V_T},\ m_u = \frac{V_o}{p_{ax}} \quad (9)$$

where p_0 is the axial sound pressure value generated at the reference point at V_T voltage, and V_o is the open-circuit voltage generated at the equivalent axial sound pressure p_{ax}. The non-axial incident sound pressure will be affected by the beampattern of the transducer's receiving array. According to the reciprocity of the transducer [11], the beampattern when transmitting is consistent with the beampattern function expression when receiving, so the axial sound pressure amplitude is equivalent to

$$p_{ax} = p_{bs}b(\theta) \quad (10)$$

According to the Equation (10), the formula for the backscattering cross-section can be expressed with the transmitting sensitivity, receiving sensitivity, beam pattern, and sampled voltage, as shown in Equation (11).

$$\sigma_{bs} = \left(\frac{V_o}{s_u m_u V_T}\right)^2 \frac{r^4 10^{-2\alpha r/10}}{b^4(\theta)} \tag{11}$$

Usually, the transducer works together with the matching circuit in the receiving state. In order to obtain the voltage in Equation (11), it is necessary to combine the transducer and its matching circuit for analysis.

The left side of Figure 2 shows a simplified circuit when the transducer is in the transmitting state [5,7,11]. The equivalent impedance of the transducer is Z_T, in which the resistance R_T of the transducer is the real part of the impedance. The active power transmitted by the circuit can be expressed as Equation (12),

$$\Pi_T = \frac{|V_T|^2}{2|Z_T|^2} R_T \tag{12}$$

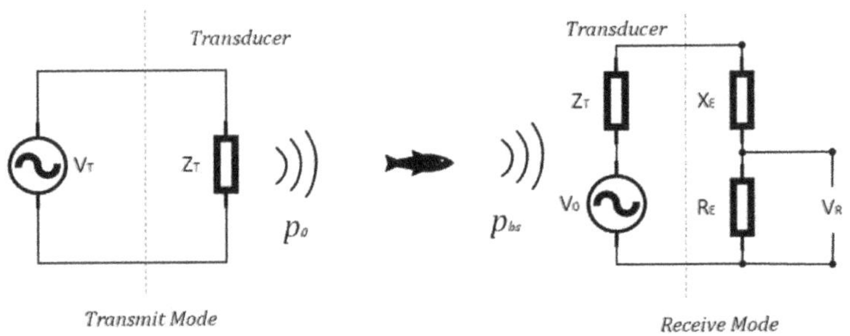

Figure 2. Helmholtz-Thevenin equivalent model for the transceiver circuit, in which the dotted line indicates the transducer. In the transmit mode, the transmit circuit can be regarded as a voltage source. In receiving mode, the voltage V_R across the resistor R_E of the matching circuit is sampled.

The transmit power can usually be measured using the power detection circuit. The amplitude of the transmit voltage obtained by deforming Equation (12) can be expressed as Equation (13).

$$V_T = \sqrt{\frac{2|Z_T|^2 \Pi_T}{R_T}} \tag{13}$$

To take the effect of non-ideal impedance into consideration when the transducer is receiving in the working state, the match circuit impedance $Z_E = R_E + iX_E$ needs to be considered as consistent [24]. The right side of Figure 2 shows the receiving equivalent circuit. The voltage collected on the equivalent resistance of the matching circuit is set to V_R. Then, the modulus value of the open-circuit voltage can be expressed as Equation (14):

$$V_o = \left|\frac{V_R}{R_E} \cdot (Z_T + Z_E)\right| \tag{14}$$

Usually, in order to ensure that the phase of the sampled electrical signal at each frequency is consistent with the phase of the received sound pressure, and to ensure that the time-domain waveform of the surface sound pressure of the transducer is consistent with the time-domain waveform of the electrical signal of the transducer under broadband signals, the matching circuit impedance Z_E is designed to minimize the impact caused by the reactance of the transducer [29]. Under ideal conditions, the parallel network of a transducer that completes impedance matching can be regarded as pure resistance R_{TE}, and its open-circuit voltage can be expressed as Equation (15).

$$V_o = \left|\frac{V_R}{R_E}\right| R_{TE} \tag{15}$$

In summary, the backscattering cross section of the target is shown as Equation (16).

$$\sigma_{bs} = \frac{|V_R|^2 R_{TE}^2 R_T}{2 s_u^2 m_u^2 |Z_T|^2 |R_E|^2 \Pi_T} \frac{r^4 10^{-2\alpha r/10}}{b^4(\theta)} \tag{16}$$

The impedance factor is defined as F_{Impe}, and the logarithms of the transmit and receive sensitivity are expressed as Su and Mu:

$$F_{Impe} = \frac{R_{TE}^2 R_T}{2|Z_T|^2 {R_E}^2} \tag{17}$$

$$Su = 20\log(s_u),\ Mu = 20\log(m_u) \tag{18}$$

Combining Equations (8) and (16)–(18), the TS can be obtained as shown in Equation (19):

$$TS = 20\log(V_R) + 10\log\left(F_{Impe}\right) - 10\log(\Pi_T) - Su - Mu + 40\log(r) + 2\alpha r - 40\log(b(\theta)) \tag{19}$$

In the actual measurement process, the impedance factor can be calculated by measuring the impedance of the transducer and the impedance of the matching circuit. The receiving sensitivity and transmitting sensitivity of the transducer are obtained by measuring the sound pressure at the acoustic axis of the specific transmitting and receiving circuit, using an oscilloscope and another transducer [30]. These parameters are usually provided in the transducer's parameter manual. The beam pattern function can be obtained by measuring the echo level of a standard metal ball at different angles [5], or by using a standard sound source [6] and fitting the obtained data through polynomial fitting.

Under a broadband signal, frequency variables can be introduced into the above equations, and considering the difference in the transducer beam pattern at different frequencies, the TS Equation (20) under broadband can be obtained,

$$\begin{aligned} TS(f) = 20\log(V_R(f)) + 10\log\left(F_{Impe}(f)\right) - 10\log(\Pi_T(f)) - Su(f) - Mu(f) \\ +40\log(r) + 2\alpha(f)r - 40\log(b(\theta,\varphi,f)) \end{aligned} \tag{20}$$

This expression only provides theoretical support for the estimation of broadband TS. The echo integration method must be further derived in the actual calculation process.

All the symbols and variables in Section 2.2 are presented in Table A2 in Appendix A, containing units and a brief description.

2.3. Echo Integration and Voltage Estimation under Broadband

Ignoring nonlinear effects and assuming that the power value and sound intensity satisfy a proportional relationship, the echo integration $[tivs]$ under narrowband signals can be defined as the integral of the square of the received voltage over time [8,26]. For the convenience of analysis, it is assumed that the echo from the target has a constant amplitude A, frequency f, and phase φ, as shown in Equation (21).

$$v(t) = A\cos(2\pi f t + \varphi) \tag{21}$$

Among these, $f = 1/T$, where T is the period of the narrowband signal. The reference resistance R_{ref} is set to 1 Ω. When the integration time τ is much larger than the period T, due to the periodicity of the cosine signal, the power Π_{ref} can be derived from the echo integration, as shown in Equation (22).

$$\Pi_{ref} = \frac{[tivs]}{\tau R_{ref}} \approx \frac{1}{TR_{ref}} \int_T v^2(t)dt = \frac{A^2}{2} \tag{22}$$

Based on the calculated power and combined with the above equation, the TS of the fish can be calculated. Compared with the original signal, the echo from fish usually shows an expansion of pulse width and a fluctuation in amplitude over time [31–33]. During the calculation of echo integration, the original signal is windowed to obtain the average TS value of the target within the time window [11].

This can then be extended to broadband signals. Assuming that there is an ideal signal, $f_1 \sim f_m$ covers m frequency points, each signal duration is T, and T satisfies $T \gg 1/f_i$ for the frequencies of these signals to meet the periodic conditions derived in Equation (22), as shown in Equation (23).

$$s_{ideal}(t) = \begin{cases} A_1 cos(2\pi f_1 t + \varphi_1), & 0 < t \leq T \\ A_2 \cos(2\pi f_2 t + \varphi_2) & T < t \leq 2T \\ \vdots & \\ A_m cos(2\pi f_m t + \varphi_m), & (m-1)T < t \leq mT \end{cases} \tag{23}$$

Regardless of the pulse width change in the echo signal, when $limT \to 0$, the weighted average of each interval integral value over time can be obtained, as shown in Equation (24).

$$\frac{[tivs]}{T} = \lim_{T \to 0} \frac{A_i^2}{T} \left(\frac{sin4\pi f_i T}{4\pi f_i} + \frac{T}{2} \right) = \frac{A_i^2}{2} \tag{24}$$

In this way, the broadband echo integral is obtained under ideal condition. However, in actual engineering applications, the time window cannot be infinitely close to zero; therefore, it seems impossible to discretely sample the signal and perform echo integration for each frequency. According to Fourier transform properties, the rotation factor $e^{-j2\pi ft}$ of the Fourier transform can be used to estimate the amplitude squared of each frequency of the all the echoes at each frequency [34]. The Fourier integral of the above signal is expressed as Equation (25).

$$S(f) = \int_{mT} s_{ideal}(t) e^{-j2\pi ft} dt \tag{25}$$

When $f \neq f_i$, the integral value is 0 due to the orthogonality of the rotation factor and the signal, and when $f = f_i$, the integral value is shown as Equation (26).

$$S(f_i) = \frac{A_i T}{2} \frac{e^{j\varphi}}{j2\pi} \tag{26}$$

Taking its modulus, we can obtain an expression for the energy meaning of A_i.

$${A_i}^2 = \frac{16\pi^2 |s(f_i) \cdot s^*(f_i)|}{T^2} \tag{27}$$

In engineering applications, an LFM signal is introduced as a broadband signal for analysis, and the expression of a typical LFM signal can be obtained in the real number domain [35], as shown in Equation (28):

$$s_{LFM} = Acos\left(2\pi f_0 t + \frac{\pi F}{T} t^2\right) = Acos(2\pi(f_0 + \frac{F}{2T}t)t) \tag{28}$$

Its equivalent frequency $f_e = f_0 + \frac{F}{2T}t$ and the duration of all signals in a narrow band can be regarded as an infinitely small value close to 0. The signals are time-varying in frequency, so the results obtained using the traditional echo integration method include the weighted average of all frequencies, which does not reflect the broadband characteristics of the target. The Fourier integral is used to estimate the amplitude of the broadband signal, and then the target broadband voltage response can be obtained.

In engineering, the received signals are sampled and quantified to obtain a discrete sequence of analog signals. Assume that the received LFM signal can be expressed as Equation (29), where A_n is the time domain response from the target.

$$x[n] = A_n\cos\left(\frac{2\pi f_n n}{f_s} + \varphi_n\right) \quad (29)$$

Since the transmitted signal is an LFM signal and the frequency is time-varying, for the convenience of analysis, the signal is subsequently converted into Fourier series form and subjected to Hilbert transform. Since the Fourier transform of LFM includes a Fresnel integral [36], the LFM frequency domain amplitude is instable, as shown in Figure 3; thus, DFT cannot be used directly to obtain the frequency domain response. In order to obtain the frequency domain response, this paper sets the ideal signal sent as the reference signal and the amplitude of the reference signal to 1 and then performs the same processing, as shown in Equations (30) and (31).

$$x_{complex}[n] = \sum_{i=1}^{N-1} A_i\exp\left(\frac{2\pi f_i n}{f_s} + \varphi_i\right) = A_n\exp\left(\frac{2\pi f_n n}{f_s} + \varphi_n\right) \quad (30)$$

$$x_{ref}[n] = \sum_{i=1}^{N-1} C_i\exp\left(\frac{2\pi f_i n}{f_s} + \varphi_i'\right) = 1\cdot\exp\left(\frac{2\pi f_n n}{f_s} + \varphi_n'\right) \quad (31)$$

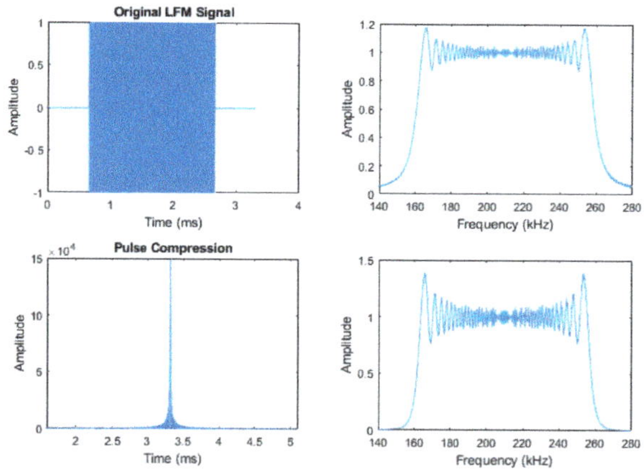

Figure 3. The upper left image shows the time domain form of the original LFM pulse, with a frequency range of 160 kHz~260 kHz, and the upper right image shows the frequency domain form of the original LFM pulse. The lower left image shows the signal after pulse compression. In order to show the changes in amplitude after pulse compression, the image was not normalized. The image on the lower right shows the spectrum after pulse compression processing. The amplitudes in the figure are reference values, without specific units.

A_i, C_i in the above formula are amplitudes in the form of a Fourier series, which are coupled with the sampling time and sampling frequency [37]; therefore, the time domain window length and sampling frequency of the received signal and the reference signal need to be consistent.

Usually, the reference signal is pulse compressed with the received signal and itself, as shown in Equations (32) and (33). This is performed to improve the signal-to-noise ratio [38]. The overall amplitude of the spectrum changes, and the pulse width in the time domain is compressed, as shown in Figure 3.

$$y_{pc} = x_{complex}[n] * conj(x_{ref}[-n]) \tag{32}$$

$$y_{ref} = x_{ref}[n] * conj(x_{ref}[-n]) \tag{33}$$

Attention: symbol * represents convolution operation in this paper.

The same reference signal is used for pulse compression. According to the convolution theorem and the principle of linear superposition [34], dividing the DFT results can eliminate the changes in spectrum amplitude caused by pulse compression. In this way, the amplitude ratio at each frequency can be obtained with Equation (34).

$$X_{responsor}(f) = \frac{DFT(y_{pc})}{DFT(y_{ref})} = \left| \sum_{i=1}^{N-1} \frac{A_i}{C_i} \delta(f - f_i) e^{j(\varphi_i - \varphi_i')} \right| \tag{34}$$

When the time domain amplitude of the reference signal is an arbitrary value, the l_2 norm of the reference signal can be used to normalize the time domain of the reference signal:

$$V_{norm}[n] = V_{ref}[n] \sqrt{\frac{N}{\left\| V_{ref}[n] \right\|_2}} \tag{35}$$

According to Equations (29)–(35), the broadband amplitude of the received voltage can be expressed by Equation (36):

$$V_R(f) = \frac{DFT\left[V_R[n] * conj\left(V_{ref}[-n]\right)\right]}{DFT\left[V_{ref}[n] * conj(V_{ref}[-n])\right]} \sqrt{\frac{\left\| V_{ref}[n] \right\|_2}{N}} \tag{36}$$

The overall process of calculating TS is summarized in Figure 4.

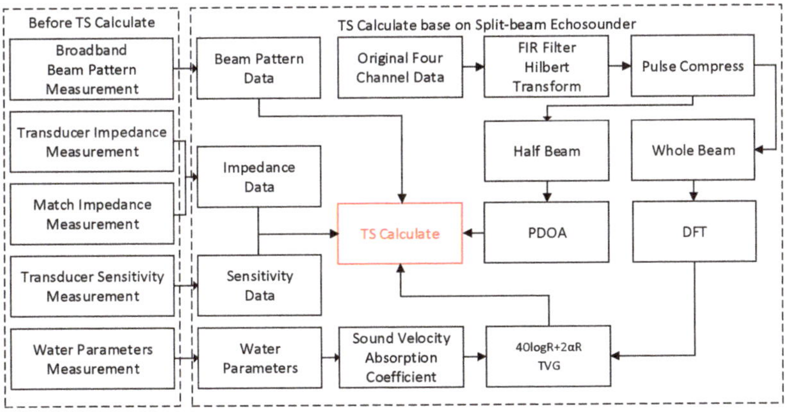

Figure 4. Overall data processing block diagram of a broadband split-beam echosounder. Before performing TS measurement, the data shown on the left side of the figure must be measured. The TS calculation includes a filtering algorithm, a positioning algorithm, and seawater absorption coefficient calculation. Due to space limitations, this is not explained in the text.

Among the calculations, the FIR filter is a commonly used noise reduction method [7,39]. The full beam is used for echo integration to obtain an estimate of TS. The phase difference of the half-beam is used to obtain the target position in the beam pattern, and the influence of the beam pattern on the full-beam TS estimation is removed [2,6,7]. In order to realize the above calculation process, experiments need to be designed to measure the above parameters.

All the symbols and variables in Section 2.3 are presented in Table A3 in Appendix A, containing units and a brief description.

3. Results

3.1. Transducer Sensitivity and Impedance Parameters

The experiment used the EK80 split beam echosounder system and selects the ES200-7C split beam broadband transducer with a nominal frequency of 200 kHz as the experimental object. The transducer manufacturer, Simrad AG (now Kongsberg Maritime AG), is located in Horten, Norway. Its parameters are shown in Table 1.

Table 1. Partial parameters of the ES200-7C transducer.

Parameters	Values
Transducer	ES200-7C
Nominal frequency (kHz)	200
Frequency range (kHz)	160~260
Beam type	Split-Beam
Nominal beam width (°)	7
Nominal Impedance (Ω)	75
Diameter (mm)	120

The sensitivity parameters and impedance under broadband were tested by the manufacturer [40], and the results are shown in Figure 5.

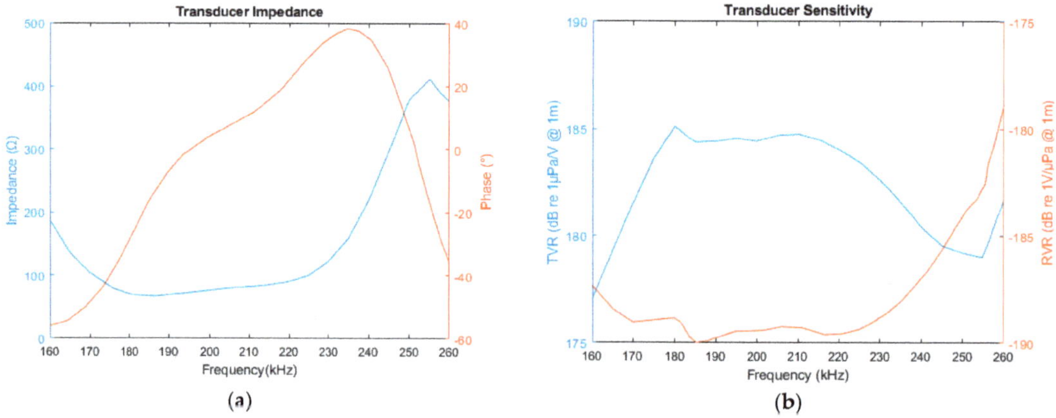

Figure 5. (**a**) The impedance of the ES200-7C transducer; (**b**) the sensitivity of the ES200-7C, which includes receiving sensitivity and transmitting sensitivity.

The impedance of the ES200-7C transducer near the nominal frequency is about 75 Ω, and the impedance is significantly greater than 75 Ω at low and high frequencies far from the center frequency. The real and imaginary reactance of the impedance vary significantly with frequency. In a circuit, resistance dissipates energy, and reactance changes the phase of the signals at different frequencies. For the transmission circuit, digital signal processing technology can be used to perform amplitude compensation or phase compensation to maintain good phase and amplitude consistency in the final transmitted signal [41].For the receiving circuit, a matching impedance network is built to reduce the impact of the nonlinear response of the transducer in the frequency domain, so that the sampled signal can be approximately proportional to the sound pressure value [29].

3.2. Matching Circuit Impedance Inversion Measurement

Since the broadband impedance data of the matching circuit inside the EK80 processor is not publicly available, a broadband sound source detection experiment must be designed to estimate the impedance of the EK80 matching circuit. This experiment was conducted in an anechoic pool with a length of 7 m, a width of 6 m, and a depth of 5 m. The sound source and standard hydrophones used the same model of hydrophones, and their parameters are shown in Table 2.

Table 2. Hydrophone parameters were used in the experiment.

Parameters	Values
Hydrophone type	Standard hydrophone
Model	TC4038-4
Effective frequency (MHz)	0.02~0.8
100 kHz receiving sensitivity (dB re 1V/μPa @1 m)	−238.09
300 kHz receiving Sensitivity (dB re 1V/μPa @1 m)	−237.81
Transmitting sensitivity (dB re 1μPa/V @1 m)	100~120
Horizontal directivity pattern (°)	360
Vertical directivity pattern (°)	120

The signal source uses DSG815, which can send AM, FM, and pulse signals in the frequency range of 9 kHz~1.5 GHz. The oscilloscope uses DS1102 with an analog bandwidth of 100 MHz, which can cover the bandwidth involved in this experiment.

The impedance can be calculated from the receiving sensitivity, sampling voltage value, and transducer impedance curve, as shown in Equation (37).

$$R_E = \frac{V_R R_T}{m_u p_r - V_R} \tag{37}$$

The parameters for the transmit signal and acquisition settings are shown in Table 3.

Table 3. Experimental emission signal parameters and oscilloscope sampling frequency.

Parameters	Values
Signal type	Linear FM
Pulse duration (ms)	1
Frequency range (kHz)	160~260
Transmitting amplitude (V)	3
Receiving amplifier gain (dB)	120
Oscilloscope sample frequency (MHz)	1

In order to estimate the equivalent impedance curve of the receiving matching circuit, the transducer was placed in the receiving state, and the EK80 system sampling voltage and standard hydrophone sound pressure were recorded. The experimental design is shown in Figure 6.

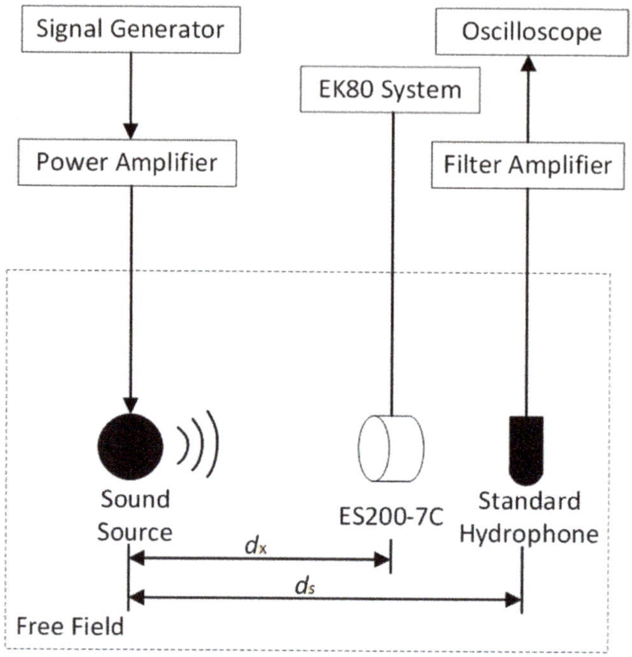

Figure 6. The dx and ds are set to 5 m, the sound source is in the axial direction of ES200-7C, and the hydrophone is placed near ES200-7C. Both are within the far-field range of the sound source. The sound pressure is kept consistent, according to the directivity of the sound source.

The sound source emits a broadband signal, and the transducer collects it using passive mode. The time domain voltage value sampled by the hydrophone is converted into the frequency domain sound pressure value $p_{ax}(f)$ using DFT and the sensitivity parameters, and then the voltage value sampled by the EK80, and the sound pressure value collected by the hydrophone are brought into the above formula. The calculated results are shown in Figure 7.

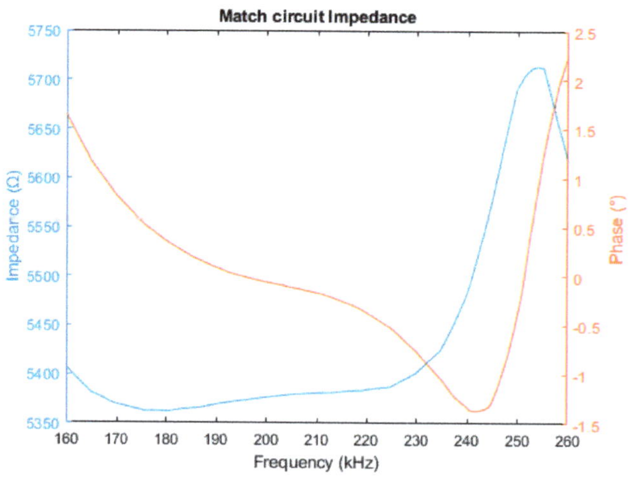

Figure 7. The matching impedance of the transducer calculated from the test data.

The impedance trend of the matching circuit is relatively consistent with that of the transducer. The changes in impedance near the frequency of 180 kHz to 220 kHz are small, and above 220 kHz the impedance changes significantly. Over most of the frequency range, the overall impedance value hovers in the range of 5350 to 5400 Ω.

3.3. Transducer Beam Pattern Measurement

Using the previous instrument, we continued to measure the beam pattern of the transducer in a broadband scenario, and the experimental design is shown in Figure 8.

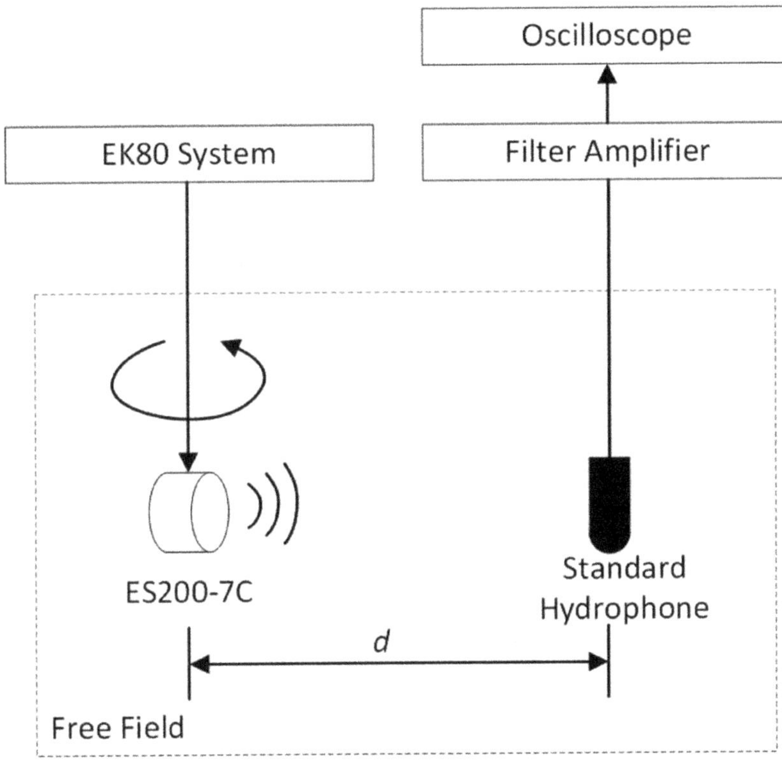

Figure 8. Beam pattern measurement experimental design. The distance d is set to 5 m to meet the far-field conditions of the transducer.

The transducer sends an LFM signal, the parameters are consistent with Table 3, and the hydrophone is used to record the sound pressure values at different angles. The received sound pressure values are subjected to discrete Fourier transform to obtain the sound pressure components at each angle. By taking the maximum point of the sound pressure ratio at the center frequency as the sound axis direction, the beam patterns at different frequencies can be obtained. The transducer is then rotated 90° and the measurement is repeated to obtain a beam pattern in the orthogonal direction.

The beam width measurement results are obtained from both directions, and the partial measurement results of the acoustic axis offset are shown in Table 4. The off-axis deviation in both directions can be calculated by the difference between the maximum value of the beam pattern and 0°, and the beam width is obtained from the difference between the left and right sides at −3 dB.

Table 4. Transducer beam pattern measurements result.

Frequency (kHz)	Alongship Beamwidth (°)	Athwartship Beamwidth (°)	Alongship Axis Bias (°)	Athwartship Axis Bias (°)
160	8.21	8.12	−0.04	0.05
165	7.82	7.76	−0.06	0.04
170	7.46	7.38	−0.05	0.06
175	7.12	7.08	−0.04	0.05
180	6.74	6.70	−0.05	0.04
185	6.47	6.46	−0.06	0.04
190	6.28	6.26	−0.06	0.06
195	6.14	6.08	−0.06	0.05
200	5.93	5.96	0	0
205	5.84	5.89	−0.04	0.02
210	5.66	5.70	0.01	0.03
215	5.46	5.58	0	0.04
220	5.40	5.51	0.02	0.02
225	5.36	5.44	0.01	0.03
230	5.30	5.31	0.01	0.03
235	5.18	5.18	0.03	0.02
240	5.10	5.13	0.03	0.02
245	5.05	5.07	0.01	0.02
250	4.98	5.01	0.04	0.04
255	4.92	4.96	0.02	0.01
260	4.85	4.92	0.03	0.02

During the measurement process, the beam pattern measurement results indicated a certain deviation in the acoustic axis. In the calculation the deviation parameter of the acoustic axis must be introduced, the measured beam pattern is polynomial fitted using the angle of two directions as independent variables, and the following beam compensation relationship is obtained [5], as shown in Equation (38), and Equation (39) describes the two variables.

$$b(\theta_{athwart}, \theta_{along}) = 10^{-0.0753*[(F_{athwart})^2+(F_{along})^2-0.18(F_{athwart})^2\cdot(F_{along})^2]} \tag{38}$$

$$F_{athwart} = \frac{|\theta_{athwart} - \theta_{athwart,bias}|}{\frac{\theta_{athwart,beamwidth}}{2}}, F_{along} = \frac{|\theta_{along} - \theta_{along,bias}|}{\frac{\theta_{along,beamwidth}}{2}} \tag{39}$$

It should be noted that to simplify the compensation of the beam pattern, spherical coordinates are not used here, but angles in two azimuths are used, and the conversion of the angles in spherical coordinates is expressed as Equation (40).

$$\begin{cases} \theta_{athwart} = arctan2(tan\theta sin\varphi) \\ \theta_{alongship} = arctan2(tan\theta cos\varphi) \end{cases} \tag{40}$$

When calculating the compensation value of the beam pattern, it is only the necessary to us the calculated azimuth angles of the target in two orthogonal directions.

3.4. Copper Ball Broadband TS Measurement

For broadband measurement, the scattering of metal balls is considered simple and ideal, and their frequency response is relatively stable in broadband. By measuring the TS of a standard scattering sphere, the rationality and effectiveness of the TS estimation algorithm can be evaluated [20]. This experiment was conducted in an anechoic pool with length, width and depth of 7 m, 6 m, and 5 m, respectively.

Before the experiment, it was necessary to measure the underwater parameters and the standard scattering sphere parameters and to set the emission signal parameters. The parameters calculated by consulting the literature are shown in Table 5 [21,42].

Table 5. Water parameters, copper ball parameters, and emission signal parameters; some parameters were directly derived from the literature or were calculated according to methods therein.

Parameter	Values
Water Parameters	
Temperature (°C)	23.2
Salinity (‰)	0
Depth (m)	2.0
Density (kg/m^3)	0.9975
Sound velocity (m/s)	1486.5
Copper Parameters	
Material	Copper
Diameter (mm)	63
Density (kg/m^3)	8947
Compressed wave speed	4760
Shearing wave speed	2288.5
Transmit Parameters	
Type	Linear FM
Pulse duration (ms)	1.024
Frequency range (kHz)	160~260
Transmit power (W)	150
Ping interval (ms)	200
Max range (m)	10

The layout of the experimental equipment is shown in Figure 9. The copper ball and transducers were, respectively, loaded onto the traveling vehicle using a rotating function. In order to verify the rationality and effectiveness of the broadband split beam TS algorithm, we set the straight-line distance between the ball and the transducer to 5 m, keeping the distance unchanged, and conducted an axial TS test. This was used to verify the performance of the algorithm without the influence of the beam pattern. The transducer was then rotated at intervals of 1° so that the copper ball was positioned at different angles of the beam pattern to verify the performance of the split-beam positioning algorithm in estimating the intensity of the nonaxial targets.

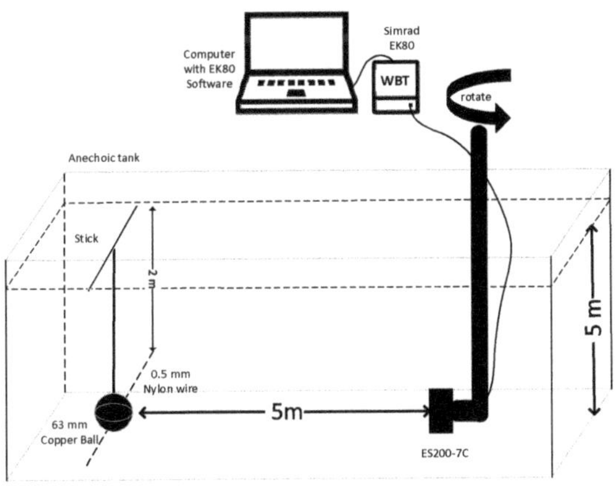

Figure 9. Experimental design for copper ball broadband TS measurement.

We used the algorithm derived in this article to directly estimate the TS of the original split-beam four-channel signal collected by the EK80, obtaining the TS data at different angles and their mean square error at different frequencies in the same image, as shown in Figure 10.

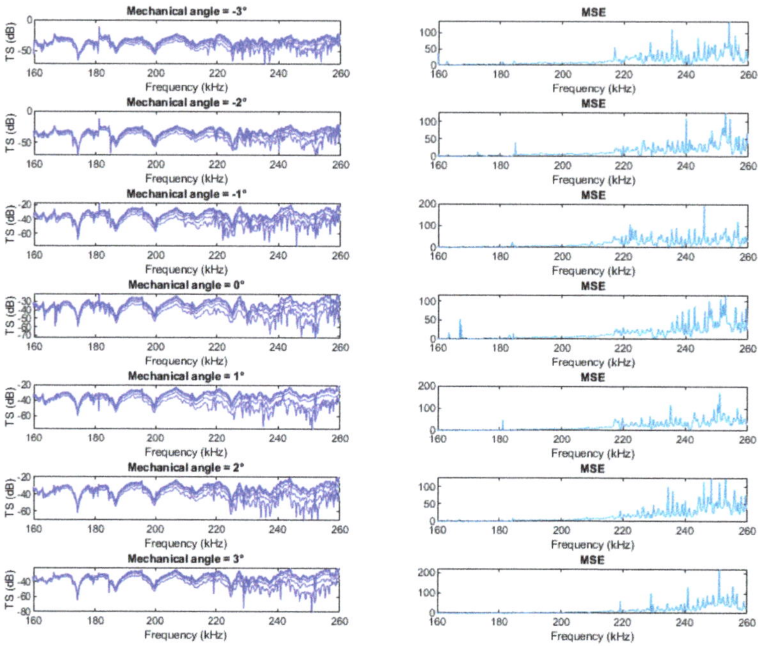

Figure 10. Broadband TS calculation and the corresponding mean square error of the copper spheres at different angles.

It was discovered that in the low-frequency range, the measured datasets were relatively stable, and the data dispersion was low. As the frequency increased, the dispersion of the measured values gradually increased. Similar situations were also reported in the literature [7,12,43].

The reason why these data fluctuate with the frequency may be explained as follows:

1. As the operating frequency of the transducer increases, the sensitivity of the transducer to changes in the surrounding environment increases [44].
2. Under water, the higher the frequency, the shorter the wavelength, and the higher the roughness of the target surface compared to the wavelength. Even for standard spheres, there may be subtle differences in scattering characteristics on different surfaces [5,20].

For point 1, it is assumed that the distribution of the coupled parameters is based on the law of large numbers and the disturbance to the signal occurs in the Gaussian distribution. Multiple pings of the original signals that meet the requirements can be weighted and averaged before broadband TS estimation is performed to obtain values that are closer to the true value [45].

For point 2, The waiting time after any movement operation during the measurement process can be extended, and the smoothness of the surface and the overall shape characteristics of the sphere to be measured can be ensured to meet the standards.

Due to the lack of professional instruments for measuring the differences in copper ball surfaces and the lack of additional tests for copper balls, this paper adopted the first method mentioned above to average the original collected signals.

The standard sphere TS can be calculated from the parameters in Table 5 [46,47]. The EK80 software also provides an estimate of TS during the data collection process. The method described in the article, the EK80 measurement data, and the theoretical TS data of the standard sphere were used for comparison. The results are shown in Figure 11.

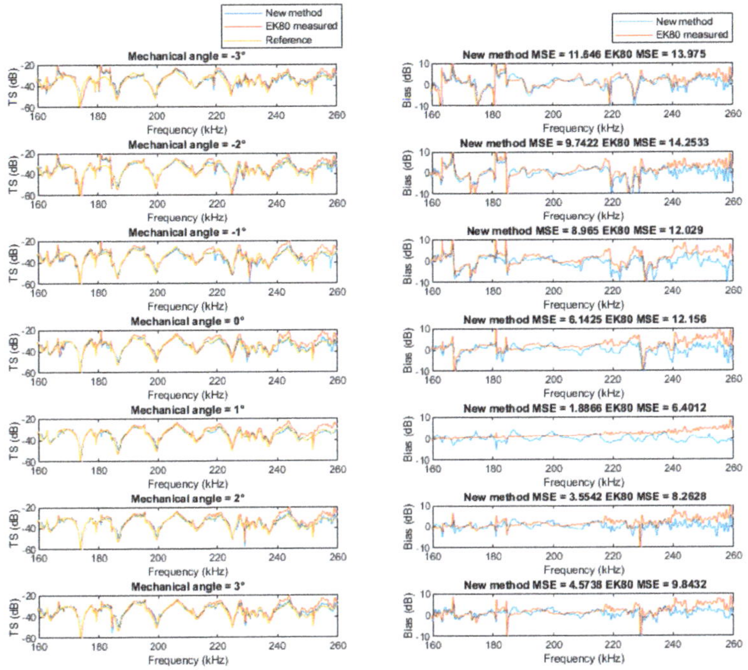

Figure 11. Comparison between the measured and reference values of copper ball broadband TS.

It can be seen that both methods show relatively consistent TS estimation performance, but there is a certain deviation from the standard value. The estimates exhibit smaller deviations in the frequencies from 190 kHz to 220 kHz. Far from the nominal frequency, the deviations are larger and exhibit more fluctuations. The mean square error of the EK80 algorithm is slightly larger than that of the algorithm mentioned in the article. This may be due to the need for recalibration as a long time has passed since the equipment was last calibrated.

4. Discussion

The broadband quantification process for echo signals is much more complicated than that under narrowband conditions, requiring the consideration of more parameters and the use of circuit analysis and complex analysis to complete quantitative analysis [7]. Compared with traditional methods, this method provides a more complete discussion of split beam technology and broadband technology. Although the steps seem more cumbersome and require professional instruments and equipment to measure various parameters of the transducer and medium parameters, the model proposed in the article can theoretically be adapted to all equipment based on split-beam technology or sonar equipment including positioning and power measurement functions. Although the variables are strictly controlled and possible disturbances in the measured values, including digital FIR filtering, matched filtering, and multiple ping data mean filtering, are taken into account, there is still a certain level of error compared to the theoretical value. The main sources of measurement errors are as follows:

1. EK80 did not provide data at the original sampling rate, but instead provided data after two decimations. The equivalent sampling frequency was only 187.5 kHz. When the pulse width is limited, the frequency resolution is low [37], and accurate frequency estimation results of voltage amplitude cannot be provided.
2. Some impedance parameters and sensitivity data are sourced from factory test reports, during long-term use, some changes may occur [5,27], resulting in errors in estimated values at some frequencies.
3. The signal propagation medium is time-varying and will be affected by the physical environment, including weather changes and vehicle shaking, causing fluctuations in the measurement results [48].
4. During long-term use the oxidation and unevenness of the surface become inconsistent with the copper ball in the reference data [20].

Regarding the first point, Jech compared two narrowband devices EK60 and EK500. Compared with EK500, the sampling rate of EK60 decreased, and the smoothness of the calculated TS also decreased [49]. This may be improved by increasing the sampling rate.

Regarding the second point, the transducer will change with the physical conditions. Dealing with time-varying nonlinear systems is complicated. The most effective and simple solution at present is to regularly re-measure the parameters mentioned in the method.

Regarding the third point, one can refer to the acquisition scheme for narrowband echo sounders provided in Ref. [48]. There is less research on broadband, which may be related to the complexity of broadband equipment.

Regarding the fourth point, chemically stable and high-hardness spheres should be used where possible. Chu et al. found that net pockets will affect the sound scattering characteristics [21]. Therefore, it is necessary to ensure that the way the sphere is suspended in the water will not affect the sound scattering characteristics. The production method of a net bag for hanging metal balls can be found in Ref. [5].

The above-mentioned errors are difficult to perfectly solve in the actual measurement process. The increase in sampling rate will incur storage and calculation problems [7]. Parameter calibration requires the use of professional instruments and specific experimental environments for measurement. Judging from the copper ball test results, the error is slightly smaller than that of the built-in algorithm of the EK80 software, which proves the rationality and effectiveness of the method proposed in this article.

From the analysis of the previous demonstration process, this broadband TS calculation method can be applied to any broadband split-beam transducer that can locate targets within the beam. In the field of fishery acoustics, a large number of studies are based on TS. The development of multi-frequency and broadband technologies in recent years has brought new possibilities to TS-based target recognition. At present, due to the particularity and complexity of broadband equipment, many problems remain to be solved before it can be fully promoted and used. It is hoped that the new method proposed in this article can be applied to diverse equipment, and that more acoustic equipment will be available in the future to provide new and valuable enhancements to fishery acoustics.

Author Contributions: Q.A. and H.L. designed the experiment and devised various testing methods; J.T. provided experimental equipment and participated in experimental planning; and J.Y. and C.L. collected data for this experiment. All authors have read and agreed to the published version of the manuscript.

Funding: This research received no external funding.

Institutional Review Board Statement: Not applicable.

Informed Consent Statement: Not applicable.

Data Availability Statement: The data that support the findings of this study are available on request from the first author, Q.A., upon reasonable request.

Acknowledgments: The authors would like to express their sincere thanks to the Institute of Hydroecology, Ministry of Water Resources and Chinese Academy of Sciences, which provided equipment for the experiment. They would also like to thank the Fishery Machinery and Instrument Research Institute. for their contribution to the experimental plan.

Conflicts of Interest: The authors declare no conflict of interest.

Appendix A

The names, symbols, and units of physical quantities must be precisely defined to ensure effective scientific communication. However, different fields of study, such as physics, electrical engineering, radar, and sonar, often use different terms and symbols, and the same symbols can be used for different items, even within the same field. In summary, scientists and researchers often exhibit preferences for specific terms and symbols. The following terminology is mainly based on the suggestions of MacLennan [50] and follows the SI units. A symbol should uniquely represent a term.

For the convenience of reading, the following table lists each variable in the formula in section order, with a brief explanation.

Table A1. Symbols, units, and descriptions for Section 2.1.

Symbol	Unit	Description
ρ_w	kg/m^3	Density of water
u_0	m/s	Particle vibration speed
a	m	Transducer diameter
k	rad/m	Wave number
c_w	m/s	Underwater sound speed
α	dB/m	Absorption coefficient
$p(r,\theta)$	Pa	The sound pressure of the transducer at any position in the field
r	m	Distance from the equivalent sound center of the transducer
p_0	Pa	Axial sound pressure amplitude at the transducer reference distance
p_{bs0}	Pa	Backscattering sound pressure at reference distance from scatterer
$b(\theta)$	None	Transducer beam pattern
p_{bs}	Pa	Scattered sound pressure at transducer reference distance
σ_{bs}	m^2	Backscattering cross section
TS	dB re 1m^2	Target strength

Table A2. Symbols, units, and descriptions for Section 2.2.

Symbol	Unit	Description
s_u	1 µPa/V@1m	Axial emission voltage sensitivity at reference distance
m_u	1V/µPa @1m	Axial received voltage sensitivity at reference distance
p_{ax}	Pa	Equivalent axial sound pressure
V_o	V	Open-circuit voltage when the transducer is receiving
V_T	V	The voltage loaded across the transducer in the transmitting state
V_R	V	The sampling voltage of the sampling circuit in the receiving state
Π_T	W	Transmit power
Z_T	Ω	Transducer impedance
R_T	Ω	Transducer resistance
Z_E	Ω	Receiver circuit impedance
R_E	Ω	Receiver circuit resistance
F_{Impe}	Ω^{-1}	Impedance factor
Su	dB re 1 µPa/V@1m	Logarithmic representation of transmitted voltage sensitivity
Mu	dB re 1V/µPa @1m	Logarithmic representation of received voltage sensitivity

Table A3. Symbols, units, and descriptions for Section 2.3.

Symbol	Unit	Description
$v(t)$	V	Assumed target echo voltage function
Π_{ref}	W	Assumed target echo power
$s_{ideal}(t)$	None	Ideal wideband signal in time domain
$S(f)$	None	Ideal wideband signal in frequency domain
t	s	Time variables in continuous time domain
T	s	Single frequency signal period
τ	s	Integration time variable
f or f_i, f_n	Hz	The signal frequency in the discrete frequency domain
F_{band}	Hz	Broadband signal bandwidth
n	None	Sequence index under time domain discretization
N	None	Finite sequence length under time domain discretization
A_n, A_i or C_i	None	Corresponding amplitude in time domain or frequency domain
x, y etc	None	Intermediate variable sequence during derivation in time domain
$X_{responsor}(f)$	None	Intermediate variable sequence during derivation in frequency domain
$V_{norm}[n]$	None	Normalized reference voltage discrete sequence
$V_{ref}[n]$	None	Reference voltage discrete sequence

References

1. Orduna Martín, C.; Encina Encina, L.; Rodríguez Ruiz, A.; Rodríguez Sánchez, V. Testing of New Sampling Methods and Estimation of Size Structure of Sea Bass (*Dicentrarchus labrax*) in Aquaculture Farms Using Horizontal Hydroacoustics. *Aquaculture* **2021**, *545*, 737242. [CrossRef]
2. Muhammad, Z.L. Acoustic Systems (Split Beam Echo Sounder) to Determine Abundance of Fish in Marine Fisheries. *Ocean. Fish.* **2017**, *3*, 555607. [CrossRef]
3. Foote, K.G.; Traynor, J.J. Comparison of Walleye Pollock Target Strength Estimates Determined from in Situ Measurements and Calculations Based on Swimbladder Form. *J. Acoust. Soc. Am.* **1988**, *83*, 9–17. [CrossRef]
4. Kang, M. Overview of the Applications of Hydroacoustic Methods in South Korea and Fish Abundance Estimation Methods. *Fish. Aquat. Sci.* **2014**, *17*, 369–376. [CrossRef]
5. Demer, D.A.; Berger, L.; Bernasconi, M.; Bethke, E.; Boswell, K.; Chu, D.; Domokos, R.; Dunford, A.; Fassler, S.; Gauthier, S. Calibration of acoustic instruments. In *ICES Cooperative Research Report*; ICES: Copenhagen, Denmark, 2015; Volume 326, 136p. [CrossRef]
6. Traynor, J.J. Fish and Standard-Sphere Targetstrength Measurements Obtained with a Dual-Beam and Split-Beam Echo Sounding System. *Rapp. P.-v. Reun. Cons. Int. Explor. Mer.* **1990**, *189*, 325–335.
7. Andersen, L.N.; Chu, D.; Heimvoll, H.; Korneliussen, R.; Macaulay, G.J.; Ona, E. Quantitative Processing of Broadband Data as Implemented in a Scientific Splitbeam Echosounder. *arXiv* **2021**, arXiv:2104.07248. [CrossRef]
8. Dragesund, O.; Olsen, S. *On the Possibility of Estimating Year-Class Strength by Measuring Echo-Abundance of 0-Group Fish*; ICES: Copenhagen, Denmark, 1965; Volume 13, pp. 44–75.
9. Jurvelius, J.; Marjomäki, T.J.; Peltonen, H.; Degtev, A.; Bergstrand, E.; Enderlein, O.; Auvinen, H. Fish Density and Target Strength Distribution of Single Fish Echoes in Varying Light Conditions with Single and Split Beam Echosounding and Trawling. *Hydrobiologia* **2016**, *780*, 113–124. [CrossRef]
10. Quazi, A. An Overview on the Time Delay Estimate in Active and Passive Systems for Target Localization. *IEEE Trans. Acoust. Speech Signal Process.* **1981**, *29*, 527–533. [CrossRef]
11. Lunde, P.; Korneliussen, R.J. Power-Budget Equations and Calibration Factors for Fish Abundance Estimation Using Scientific Echo Sounder and Sonar Systems. *J. Mar. Sci. Eng.* **2016**, *4*, 43. [CrossRef]
12. Demer, D.A.; Andersen, L.N.; Bassett, C.; Berger, L.; Chu, D.; Condiotty, J.; Hutton, B.; Korneliussen, R.; Bouffant, N.L.; Macaulay, G. *2016 USA–Norway EK80 Workshop Report: Evaluation of a Wideband Echosounder for Fisheries and Marine Ecosystem Science*; ICES Cooperative Research Reports (CRR); ICES: Copenhagen, Denmark, 2017; Volume 336. [CrossRef]
13. Rautureau, C.; Goulon, C.; Guillard, J. In Situ TS Detections Using Two Generations of Echo-Sounder, EK60 and EK80: The Continuity of Fishery Acoustic Data in Lakes. *Fish. Res.* **2022**, *249*, 106237. [CrossRef]
14. Simrad EK500 Operator Manual. Available online: https://www.kongsberg.com/maritime/support/simradsupport/simrad-docs/discontinued-echo-sounders/simrad-ek500/ (accessed on 21 December 2023).
15. Ishimaru, A. Theory and Application of Wave Propagation and Scattering in Random Media. *Proc. IEEE* **1977**, *65*, 1030–1061. [CrossRef]
16. Hao, C.; Orlando, D.; Liu, J.; Yin, C. Introduction to Radar Systems. In *Advances in Adaptive Radar Detection and Range Estimation*; Springer: Berlin/Heidelberg, Germany, 2022; ISBN 9789811663987.

17. Bjørnø, L. Chapter 5—Scattering of Sound. In *Applied Underwater Acoustics*; Neighbors, T.H., Bradley, D., Eds.; Elsevier: Amsterdam, The Netherlands, 2017; pp. 297–362.
18. Yang, Y.; Gastauer, S.; Proud, R.; Mangeni-Sande, R.; Everson, I.; Kayanda, R.J.; Brierley, A.S. Modelling and in Situ Observation of Broadband Acoustic Scattering from the Silver Cyprinid (*Rastrineobola argentea*) in Lake Victoria, East Africa. *ICES J. Mar. Sci.* **2023**. [CrossRef]
19. Stanton, T.K.; Chu, D. Calibration of Broadband Active Acoustic Systems Using a Single Standard Spherical Target. *J. Acoust. Soc. Am.* **2008**, *124*, 128–136. [CrossRef] [PubMed]
20. Hobaek, H.; Forland, T.N. Characterization of Target Spheres for Broad-Band Calibration of Acoustic Systems. *Acta Acust. United Acust.* **2013**, *99*, 465–476. [CrossRef]
21. Chu, D.; Eastland, G.C. Calibration of a Broadband Acoustic Transducer with a Standard Spherical Target in the near Field. *J. Acoust. Soc. Am.* **2015**, *137*, 2148–2157. [CrossRef] [PubMed]
22. Lavery, A.C.; Chu, D.; Moum, J.N. Measurements of Acoustic Scattering from Zooplankton and Oceanic Microstructure Using a Broadband Echosounder. *ICES J. Mar. Sci.* **2010**, *67*, 379–394. [CrossRef]
23. Butler, J.L.; Sherman, C.H. Chapter 5—Transducers as Projectors. In *Transducers and Arrays for Underwater Sound Compress*; Springer: Berlin/Heidelberg, Germany, 2016; pp. 185–272.
24. Lunde, P. Finite-Amplitude Power Budget Equations for Acoustic Fish Abundance Estimation. *J. Mar. Sci. Eng.* **2020**, *8*, 98. [CrossRef]
25. Butler, J.L.; Sherman, C.H. Chapter 10—Acoustic Radiation from Transducers. In *Transducers and Arrays for Underwater Sound Compress*; Springer: Berlin/Heidelberg, Germany, 2016; pp. 517–551.
26. Medwin, H.; Clay, C.S. Chapter 7—Sound Scattered by a Body. In *Fundamentals of Acoustical Oceanography*; Medwin, H., Clay, C.S., Eds.; Applications of Modern Acoustics; Academic Press: San Diego, CA, USA, 1998; pp. 234–286.
27. Demer, D.A.; Renfree, J.S. Variations in Echosounder–Transducer Performance with Water Temperature. *ICES J. Mar. Sci.* **2008**, *65*, 1021–1035. [CrossRef]
28. Butler, J.L.; Sherman, C.H. Chapter 6—Transducers as Hydrophones. In *Transducers and Arrays for Underwater Sound Compress*; Springer: Berlin/Heidelberg, Germany, 2016; pp. 281–341.
29. Wang, Y.; Sun, D.; Yong, J. Design of Broadband Matching Circuit for Underwater Acoustic Communication Transducer. In Proceedings of the 2015 International Conference on Intelligent Systems Research and Mechatronics Engineering, Zhengzhou, China, 11–13 April 2015; Atlantis Press: Amsterdam, The Netherlands, 2015; pp. 2298–2305.
30. Xiao, D.; Fan, Q.; Xu, C.; Zhang, X. Measurement Methods of Ultrasonic Transducer Sensitivity. *Ultrasonics* **2016**, *68*, 150–154. [CrossRef]
31. Soliveres, E.; Poveda, P.; Estruch, V.D.; Pérez-Arjona, I.; Puig, V.; Ordóñez, P.; Ramis, J.; Espinosa, V. Monitoring Fish Weight Using Pulse-Echo Waveform Metrics. *Aquac. Eng.* **2017**, *77*, 125–131. [CrossRef]
32. Puig-Pons, V.; Muñoz-Benavent, P.; Pérez-Arjona, I.; Ladino, A.; Llorens-Escrich, S.; Andreu-García, G.; Valiente-González, J.M.; Atienza-Vanacloig, V.; Ordoñez-Cebrian, P.; Pastor-Gimeno, J.I. Estimation of Bluefin Tuna (*Thunnus thynnus*) Mean Length in Sea Cages by Acoustical Means. *Appl. Acoust.* **2022**, *197*, 108960. [CrossRef]
33. Kubilius, R.; Bergès, B.; Macaulay, G.J. Remote Acoustic Sizing of Tethered Fish Using Broadband Acoustics. *Fish. Res.* **2023**, *260*, 106585. [CrossRef]
34. Oppenheim, A.V.; Willsky, A.S.; Nawab, S.H.; Ding, J.-J. Chapter 3—Fourier Series Representation of Periodic Signals. In *Signals and Systems*; Prentice Hall: Upper Saddle River, NJ, USA, 1997; pp. 177–250.
35. Liu, D.; Qu, H.; Wang, W.; Deng, J. Multiple Targets Detection of Linear Frequency-Modulated Continuous Wave Active Sonar Using Fractional Fourier Transform. *Integr. Ferroelectr.* **2020**, *209*, 1–10. [CrossRef]
36. Guo, Y.; Yang, L. Method for Parameter Estimation of LFM Signal and Its Application. *IET Signal Process.* **2019**, *13*, 538–543. [CrossRef]
37. Oppenheim, A.V.; Willsky, A.S.; Nawab, S.H.; Ding, J.-J. Chapter 5—The Discrete-Time Fourier Transform. In *Signals and Systems*; Prentice Hall: Upper Saddle River, NJ, USA, 1997; pp. 358–396.
38. Levanon, N.; Mozeson, E. Chapter 2—Matched Filter. In *Radar Signals*; John Wiley & Sons: Hoboken, NJ, USA, 2004; pp. 20–33.
39. Oppenheim, A.V.; Willsky, A.S.; Nawab, S.H.; Ding, J.-J. Chapter 6—Time and Frequency Characterization of Signals and Systems. In *Signals and Systems-6*; Prentice Hall: Upper Saddle River, NJ, USA, 1997; pp. 423–472.
40. Simrad ES200-7C. Available online: https://www.kongsberg.com/maritime/products/commercial-fisheries/td/200-khz/simrad-es200-7c/ (accessed on 20 December 2023).
41. Li, J.; Jiang, L.; Yu, F.; Zhang, Y.; Gao, K. Research on Improving Measurement Accuracy of Acoustic Transfer Function of Underwater Vehicle. In *MATEC Web of Conferences*; EDP Sciences: Les Ulis, France, 2021; Volume 336, p. 01006. [CrossRef]
42. Ainslie, M.A.; McColm, J.G. A Simplified Formula for Viscous and Chemical Absorption in Sea Water. *J. Acoust. Soc. Am.* **1998**, *103*, 1671–1672. [CrossRef]
43. Conti, S.G.; Demer, D.A. Wide-Bandwidth Acoustical Characterization of Anchovy and Sardine from Reverberation Measurements in an Echoic Tank. *ICES J. Mar. Sci.* **2003**, *60*, 617–624. [CrossRef]
44. Au, W.W.; Benoit-Bird, K.J. Broadband Backscatter from Individual Hawaiian Mesopelagic Boundary Community Animals with Implications for Spinner Dolphin Foraging. *J. Acoust. Soc. Am.* **2008**, *123*, 2884–2894. [CrossRef]

45. De Robertis, A.; Higginbottom, I. A Post-Processing Technique to Estimate the Signal-to-Noise Ratio and Remove Echosounder Background Noise. *ICES J. Mar. Sci.* **2007**, *64*, 1282–1291. [CrossRef]
46. Demer, D.A.; Conti, S.G.; De Rosny, J.; Roux, P. Absolute Measurements of Total Target Strength from Reverberation in a Cavity. *J. Acoust. Soc. Am.* **2003**, *113*, 1387–1394. [CrossRef]
47. Maclennan, D.N. The Theory of Solid Spheres as Sonar Calibration Targets. *Scottish Fish. Res. Rep.* **1981**, *22*, 1–16.
48. Jech, J.M.; Schaber, M.; Cox, M.; Escobar-Flores, P.; Gastauer, S.; Haris, K.; Horne, J.; Jarvis, T.; Ladroit, Y.; O'Driscoll, R. Collecting Quality Echosounder Data in Inclement Weather. In *ICES Cooperative Research Report*; ICES: Copenhagen, Denmark, 2021; Volume 352, 108p. [CrossRef]
49. Jech, J.M.; Foote, K.G.; Chu, D.; Hufnagle Jr, L.C. Comparing Two 38-kHz Scientific Echosounders. *ICES J. Mar. Sci.* **2005**, *62*, 1168–1179. [CrossRef]
50. MacLennan, D.N.; Fernandes, P.G.; Dalen, J. A Consistent Approach to Definitions and Symbols in Fisheries Acoustics. *ICES J. Mar. Sci.* **2002**, *59*, 365–369. [CrossRef]

fishes

MDPI

Article

Spatiotemporal Distribution of Antarctic Silverfish in the Ross Sea, Antarctica

Sara Lee [1], Wooseok Oh [2], Hyoung Sul La [3], Wuju Son [3,4], Jeong-Hoon Kim [5] and Kyounghoon Lee [6,*]

1 Department of Fisheries Physics, Pukyong National University, Busan 48513, Republic of Korea; wogurtlfgdj@naver.com
2 Institute of Low-Carbon Marine Production Technology, Pukyong National University, Busan 48513, Republic of Korea; owsnice@gmail.com
3 Division of Ocean & Atmosphere Sciences, Korea Polar Research Institute, Incheon 21990, Republic of Korea; hsla@kopri.re.kr (H.S.L.); swj5753@kopri.re.kr (W.S.)
4 Department of Polar Sciences, University of Science and Technology, Daejeon 34113, Republic of Korea
5 Division of Life Sciences, Korea Polar Research Institute, Incheon 21990, Republic of Korea
6 Division of Marine Production System Management, Pukyong National University, Busan 48513, Republic of Korea
* Correspondence: klee71@pknu.ac.kr; Tel.: +82-51-629-5889; Fax: +82-51-629-5886

Abstract: Antarctic silverfish (*Pleuragramma antarcticum*) play a crucial intermediary role in connecting top predators and krill in the food web of the Antarctic Ocean. Despite their crucial role, research on their abundance is lacking. In this study, we estimated the abundance of juvenile Antarctic silverfish as foundational data for predicting their abundance. The density of juvenile Antarctic silverfish was estimated using an acoustic backscattering theoretical model. The mean volume backscattering strength was used to investigate the vertical and horizontal distributions of juvenile Antarctic silverfish in the Antarctic Ross Sea. The survey area was located near Cape Hallett, Antarctica, where Antarctic krill (*Euphausia superba*), ice krill (*E. crystallorophias*), and Antarctic silverfish coexist. The survey was performed four times using the Korean Antarctic research ship, RV Araon (R/V, 7507 GT). Frame trawls were conducted to identify the length and weight of the target fish species in the survey area. Captured Antarctic silverfish captured measured 3–9 cm. The maximum target strength (TS) was −92.93 dB at 38 kHz, −86.63 dB at 120 kHz, and 85.89 dB at 200 kHz. The average TS was −100.00 dB at 38 kHz, −93.00 dB at 120 kHz, and −106.90 dB at 200 kHz. Most juvenile Antarctic silverfish were found at a depth of 100 m and were distributed closer to sea ice. Between nearshore and polynya waters, the fish demonstrated a proclivity for polynya waters.

Keywords: Antarctic silverfish; hydroacoustic; mean volume backscattering strength

Key Contribution: Antarctic silverfish play a crucial intermediary role in the food web of the Antarctic ecosystem, making it a highly important species. Assessing the resources of Antarctic silverfish is vital, particularly using underwater acoustics as a technique. Target strength (TS) is a crucial factor in evaluating resources through acoustics, and we estimated the TS of Antarctic silverfish, elucidating their spatiotemporal distribution. This paper contributes to a better understanding of the Antarctic ecosystem.

1. Introduction

Antarctic silverfish (*Pleuragramma antarcticum*) is an important fish species in the Antarctic Ross Sea owing to its wide distribution and high abundance, serving as prey for both marine mammals and birds [1]. It also plays an important role in the food chain of the Ross Sea by feeding on Antarctic krill (*Euphausia superba*), linking top predators to krill [2–4]. Therefore, Antarctic silverfish have a significant impact on the abundance of top predators and krill in the Antarctic.

Antarctic silverfish is the only notothenioid species with an entire pelagic life cycle [5]. It also has a larval stage of more than one year, which is longer than that of other fish species. The larvae are initially distributed in surface waters, and as they mature, they disperse into deeper waters [6,7].

Antarctic silverfish can be classified into three age groups based on their length [3,8,9]. Specimens between 0.8 and 3 cm in length are classified as post-larvae, and those between 3 and 10 cm in length as juvenile fish. Specimens larger than 11 cm in length are classified as adult fish. Juvenile Antarctic silverfish is highly abundant in Antarctic waters, accounting for 98% of the total plankton [10].

The abundance of a target species is generally predicted based on the abundance of its juveniles. Therefore, estimating the abundance of juvenile Antarctic silverfish is a crucial factor in predicting the overall abundance of Antarctic silverfish.

Among various methods used to determine the density and distribution of organisms, the hydroacoustic technique is the most effective method for surveying waters with limited access and time limits, like those in Antarctica. To extract echo signals from the target fish species using the hydroacoustic technique, it is important to discern the target strength (TS) of the target fish species. The acoustic scattering model is most commonly used to measure the TS of fish species, as this model takes into account several variables, such as size, swimming angle, size of the swimbladder, and usage frequency, as well as morphological characteristics of the target fish species. In the acoustic scattering theoretical model (Kirchhoff–Ray mode), TS can be estimated by calculating the sum of the volumes of the fish body and swimbladder, approximated as cylindrical shapes. This theoretical model is a method of estimating acoustic backscatter strength by separating specific anatomical structures of the target species, which is generally more precise than estimating strength using only the organism's shape [11].

Abundance estimation of a target species using acoustic techniques can be calculated by dividing the volume backscattering strength (Sv) collected in the field by the target strength (TS) of the target species. However, given the diversity of species in the ocean, it is essential to first identify the target species and then separate their signals for accurate analysis. There are two methods for separating the signals of the target species: time-varied threshold (TVT), which uses one frequency, and mean volume backscattering strength (MVBS), which uses more than two frequencies [12].

In this study, we measured the TS of juvenile Antarctic silverfish using an acoustic backscattering theoretical model to estimate the spatiotemporal distribution of juvenile Antarctic silverfish. We utilized MVBS to determine the vertical and horizontal distributions of juvenile Antarctic silverfish in the Antarctic Ross Sea. We believe this study can provide foundational data for predicting the abundance of Antarctic silverfish.

2. Materials and Methods

2.1. Survey Regions and Acoustic Data Collection

The study area was in the waters near Cape Hallet, Antarctica, where Antarctic krill (*E. superba*), ice krill (*E. crystallorophias*), and Antarctic silverfish coexist. Furthermore, the survey area is a known nursery ground for Antarctic silverfish [13]. The survey was conducted four times: from 16 February to 10 March 2018; 21 December 2018 to 18 January 2019; 3 March to 7 April 2020; and 6 December to 22 December 2020 (Table 1). The acoustic transects in each survey were 6, 12, 12, and 10, respectively. The Korean Antarctic research ship, the icebreaker RV Araon (R/V, 7507 G/T), was used for the survey. Acoustic data were collected while maintaining a vessel speed of 7–10 knots (Figures 1 and 2).

Table 1. Survey period.

Survey Rounds	Survey Period	Survey Transects
1st	16 February 2018–10 March 2018	06
2nd	21 December 2018–18 January 2019	12
3rd	3 March 2020–7 April 2020	12
4th	6 December 2020–22 December 2020	10

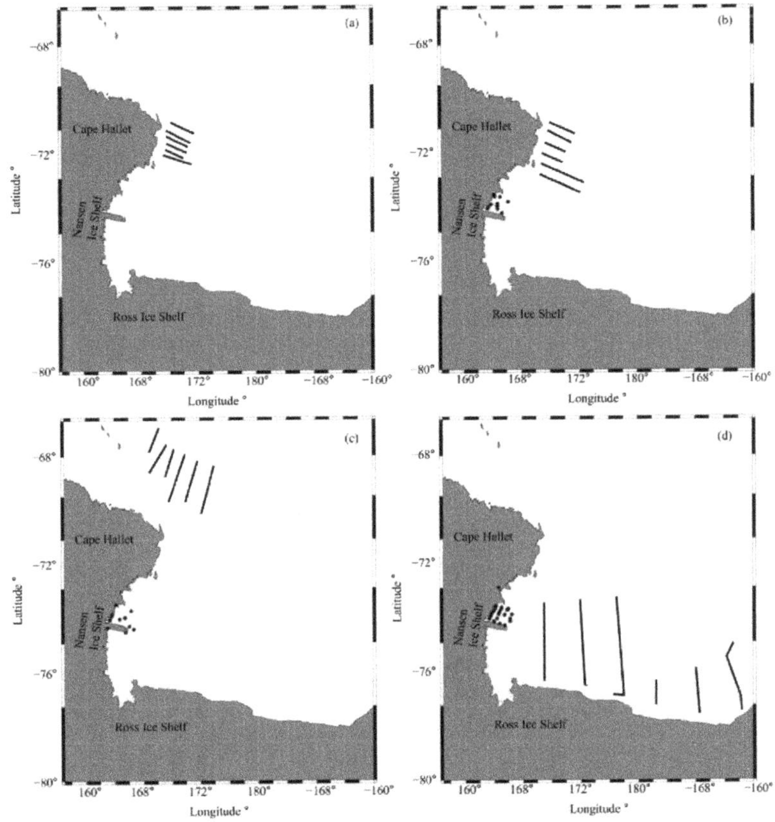

Figure 1. Location map of the acoustic survey transects in the Ross Sea. (**a**) February 2018, (**b**) December 2018, (**c**) March 2020, and (**d**) December 2020. The black dots represent icons generated for drawing lines, while the lines symbolize acoustic survey transects.

2.2. Sample Catch Data

It was necessary to determine the length and weight of the target fish species to evaluate species density via hydroacoustics. Therefore, quantitative fishing gear to collect the target fish species in the survey area was required. Since the shape of the net structure changes depending on the speed of the survey vessel, a small midwater trawl with a frame trawl of 2 m × 2 m was used. The frame was made of stainless steel with a diameter of 100 mm. For stable fishing gear deployment, five buoys were installed on the top for buoyancy, and four 100 kg weights were attached to the lower bar for reinforcing weight and stability (Figure 3). The net was manufactured in duplicate by using double-walled nets of approximately 7 m; the outer net protected the inner net from breakage. The mesh sizes of the inner and outer nets were 5 and 10 mm, respectively. The square frame net was used as fishing gear to collect fish larvae and specimens.

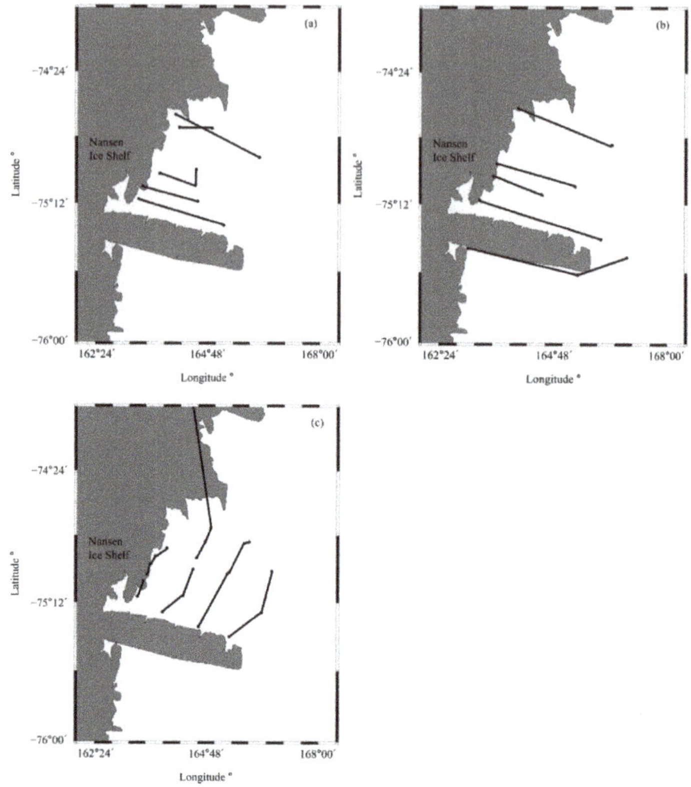

Figure 2. Location map of the acoustic survey transects in polynya waters. (**a**) December 2018, (**b**) March 2020, and (**c**) December 2020.

Figure 3. Frame midwater trawl and attached buoys.

Collections via the frame trawl were carried out on the Ross Sea ice shelf, located offshore of the continental shelf in relatively shallow water with a depth ranging from 600 to 700 m [14]. A total of eight voyages were carried out between 7 and 16 December 2020 (Figure 4). While maintaining a vessel speed of 2–3 knots, the net was towed for more than 30 min, and acoustic data were collected simultaneously.

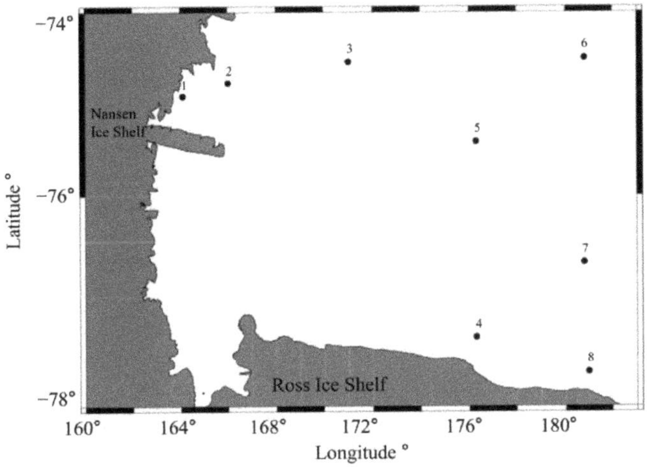

Figure 4. Location map of the frame midwater trawl stations conducted in the Ross Sea.

2.3. The Kirchhoff–Ray Mode (KRM)

The Kirchhoff–Ray mode (KRM) model was applied to estimate the acoustic scattering intensity of Antarctic silverfish in the Antarctic Ross Sea. The KRM model (Equations (1) and (2)) can identify and quantify the shape of the fish body to estimate acoustic scattering strength [15].

$$L_{FISH} = L_{BODY} = f(f_r, \theta_{tilt}, S_b, \rho_w, \rho_b, c_w, c_b) \quad (1)$$

$$TS = 10\log_{10}|L_{FISH}|^2 \quad (2)$$

where f_r is the frequency, θ_{tilt} is the angle between the body axis and the incidence angle, S_b is the body shape of the fish approximated as a cone, ρ_w is the density of the medium (seawater), ρ_b is the density of the fish body, c_w is the velocity of sound of the medium (seawater), and c_b is the velocity of sound of the fish body.

The TS using the KRM model calculates the posture angle from $-60°$ to $60°$ at $1°$ intervals. The maximum and average values of TS were evaluated; the average values (Equations (3) and (4)) were calculated using the probability density function (PDF), while assuming an average posture angle and standard deviation of $-5°$ and $15°$, respectively, for common fish. The TS of each posture angle calculated every $1°$ was replaced by the scattering cross-section, multiplied by the PDF of the posture angle of $-5 \pm 15°$, and then the average TS was calculated using the sum [16]:

$$\sigma_{bs} = \int_{-\pi/2}^{\pi/2} \sigma(\theta) f(\theta) d\theta \quad (3)$$

$$TS_{avg.} = 10\log_{10}\sigma_{avg.} \quad (4)$$

where $\sigma(\theta)$ is the backscattering cross-section at each swimming angle θ, and $f(\theta)$ refers to the frequency of occurrence of each swimming angle. Additionally, the TS relation for Antarctic silverfish can be expressed by Equations (5) and (6). Equation (6) assumes that the reflection intensity is proportional to the second power of the length:

$$TS = a\log_{10} L + b \quad (5)$$

$$TS = 20log_{10}L + TS_{10} \quad (6)$$

where a is a slope, b is an intercept, and L indicates length (cm).

The KRM model estimates acoustic scattering strength by calculating the sum of the approximate volumes of the fish body and the swimbladder. However, in the case of Antarctic silverfish, the formula for the swimbladder was removed when applying the model, as this species does not have a swimbladder [13]. For the density and sound velocity ratios of the target species, this study utilized the findings of a previous study, which were 1.012 and 1.015, respectively [17].

Since contour data are required to use the KRM model, photographs were taken by enumerating Antarctic silverfish. The pictures of the samples were analyzed using a digitizing software program (Getdata Ver. 2.26, https://getdata-graph-digitizer.software.informer.com; accessed on 14 December 2023) to collect body shape coordinates by dividing the sides of the fish body into 0.2 mm intervals. This information was applied in the KRM model.

2.4. Data Processing

Acoustic data acquired through a scientific echosounder often include various acoustic noises, leading to the degradation of the normal echo signal. The noise in the acoustic data of this study included background, transient, and impulse noises (Figure 5). Background noise occurs when the survey depth is out of the detection range of the used frequency, and the acoustic intensity is relatively strong. Noise was removed by considering the values of the data samples in the horizontal and vertical ranges. Background noise was removed by converting the value of the data sample to −999 dB if the signal-to-noise ratio (SNR) value was greater than the threshold value(δ) when artificial noise was subtracted from the data sample in the specified area; this method is called TVT (Table 2). The SNR at the ping (i) and range sample (j) is defined as follows:

$$SNR(i, j) = S_{v,\,corr}(i, j) - S_{noise}(i, j)$$

$$if\ SNR(i, j) \leq threshold_{SNR'} \quad (7)$$

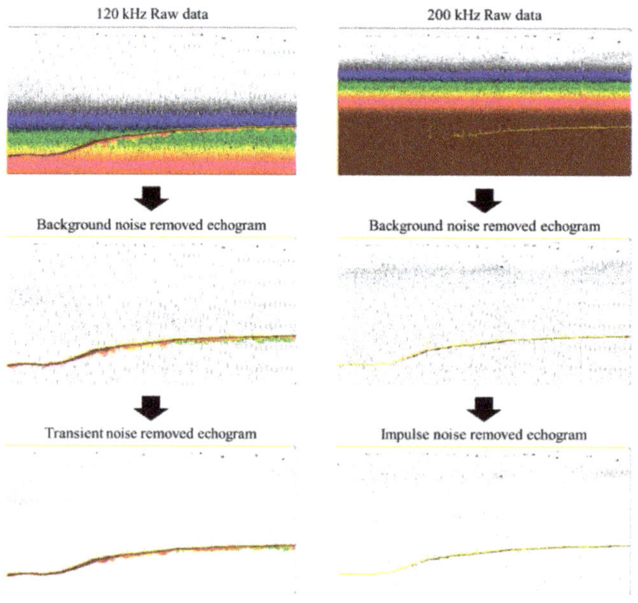

Figure 5. Examples of noise processing in the echogram. The intensity is indicated by the color, with red representing high intensity and blue indicating low intensity.

Table 2. Background noise parameters.

Frequency	120 kHz	200 kHz
Horizontal extent (ping)	20	20
Vertical units	samples	samples
Vertical extent (samples)	5	5
Vertical overlap (%)	0	0
Maximum noise (dB)	−125	−250
Minimum SNR	10	10

Transient noise is electrical noise caused by various on-board electrical equipment and is characterized by a regular drizzling pattern [18]. Transient noise was removed as follows: when a range of ~7 pings in the horizontal direction of the data sample was set, and the median value of the area was subtracted from the center data sample in the specified area, the noise was removed if the value was greater than the threshold value (Table 3).

$$S_{vi,j} - \overset{\sim}{S_{vm,n}} > \delta. \tag{8}$$

Table 3. Transient noise parameters.

Frequency	120 kHz
Exclude above	Fixed depth surface exclusion at 10 m
Exclude below	Minimum integration stop depth
Exclude below threshold (dB at 1 m)	−170
Vertical window units	Samples
Vertical window size (samples)	5
Horizontal size (ping)	7
Vertical size (samples)	9
Calculations per sample	63
Percentile	50
Threshold (dB)	10
Noise sample replacement value	Percentile
percentile	10

Impulse noise occurs due to interferences from acoustic equipment installed on other ships and is characterized by irregular, thick rain patterns. Impulse noise was removed using the following process: after each sample value of the horizontal range variations was subtracted from the central data sample of the specified area, a value was obtained. If the value was greater than the threshold value, the noise was removed; this method is called the two-sided comparison method (Table 4).

$$S_{vi,j} - S_{v(i+n),\quad j} > \delta and$$

$$S_{vi,j} - S_{v(i-n),\quad j} > \delta \tag{9}$$

Table 4. Impulse noise parameters.

Frequency	200 kHz
Exclude above	Surface
Exclude below	Bottom
Exclude below threshold (dB at 1 m)	−150
Vertical window units	Samples
Vertical window size (samples)	3
Horizontal size (pings)	3
Threshold (dB)	10
Noise sample replacement value	Mean

The frequency difference was the difference in the MVBS. The acoustic backscatter strength of the target species was extracted to obtain a positive value; low frequencies were subtracted from high frequencies. The acoustic scattering strength of the scatter predicted by the KRM model was calculated to be −92.69 dB at 120 kHz and −90.46 dB at 200 kHz when considering the minimum length (3 cm), and −69.95 dB at 120 kHz and −70.72 dB at 200 kHz when considering the maximum length (9 cm). Therefore, the $MVBS_{200kHz-120kHz}$ range of Antarctic silverfish in the echograms of the two frequencies received from the same seawater volume was from 2.23 to −0.77 dB. The frequency difference was applied to the results of the collections in December 2020. The acoustic backscatter strength of Antarctic silverfish was greater at 200 kHz than at 120 kHz.

To calculate the $MVBS_{200kHz-120kHz}$ within the same seawater volume after removing the noise from the sea surface and seafloor, the cell size (width × length) per frequency was applied to 1 ping × 1 m and integrated to generate a new echogram. A data range bitmap that set the frequency difference range of Antarctic silverfish was created. A mask operator was used to separate the echo signal from Antarctic silverfish by superimposing the echoes that matched the cell size with 120 kHz frequency at 200 kHz. Matched echoes were considered to be fish.

To estimate the density of juvenile Antarctic silverfish using acoustics, volume backscattering strength (*Sv*) data extracted from the scientific echosounder at 1 n.mile intervals were converted to nautical area scattering coefficients (*NASCs*). The relationship for converting *Sv* to *NASC* can be found in Equation (10):

$$NASC = 4\pi 1852^2 \int_{r_1}^{r_2} Svdr \tag{10}$$

Since the *NASC* value is the linear sum of the signals received from aquatic organisms in the water volume, the density of the target organisms (ρ, g/m^2) can be calculated by dividing the average *NASC* value in the obtained seawater volume by the *TS* of the target fish; the *TS* and backscattering cross-section according to the length (*L*, mm) of the target organisms can be expressed in Equations (12) and (13), respectively:

$$NASC = \rho \cdot TS \tag{11}$$

$$TS = 20log(L) + TS_{mm} \tag{12}$$

$$\sigma = 4\pi 10^{\frac{TS}{10}} \tag{13}$$

Additionally, the length (*L*, mm)–weight (*w*, mg) relationship of the target organism can be found in Equation (14):

$$\omega = \alpha L^b \tag{14}$$

Here, the backscattering cross-section and length–weight function of the target organism were obtained from the catch data collected at the same time as the acoustic survey. The density of the target species (ρ) (Equation (15)) can be calculated by dividing the average *NASC* within the seawater volume at 1 n.mile intervals by the backscattering cross-section (σ) of the target species and multiplying it by its weight. The remainder on the right side of Equation (15), except for *NASC*, is the conversion factor (CF) that calculates density from acoustic data, considering the backscattering cross-section and length–weight of the target species. Average values were utilized for the backscattering cross-section and weight of the target species.

$$\rho = \left(\frac{NASC}{\sigma}\right) \cdot \omega = \frac{\alpha L^b}{4\pi 10^{TS/10}} \cdot NASC \tag{15}$$

The average target species density ($\overline{\rho}$) of the entire survey area represents the weighted mean of the average density data per vessel, as shown in Equation (16):

$$\overline{\rho} = \frac{\sum_{i=1}^{N} \overline{\rho} \cdot n_i}{\sum_{i=1}^{N} n_i} \tag{16}$$

where $\overline{\rho}_i$ is the mean density of the ith vessel, n_i is the number of the ith vessel (elementary distance sampling unit, EDSU), and N indicates the number of vessels.

3. Results

A distribution survey of Antarctic silverfish was performed using a frame trawl in the Antarctic Ross Sea in December 2020. The individuals captured in the fishing gear were juvenile Antarctic silverfish. Antarctic silverfish were most abundantly caught at station 5, with a total of 46 individuals. (Table 5). The catch ratio of Antarctic silverfish was >53% at station 1 (Figure 6). The variation in the lengths of the fish obtained using the fishing gear is shown in Figure 6. Antarctic silverfish caught had a length of 3–9 cm. Fish with a 5 cm body length were found in the highest proportion, accounting for approximately 53% of the captured fish, whereas 6 cm fish were captured in the lowest proportion (Figure 7).

Table 5. Trawl time, station, location, number caught, and catch rate.

Station	Date (DD.MM.YYYY)	Latitude (°)	Longitude (°)	Number Caught (N)	Antarctic Silverfish Ratio (% by Number)
1	12.07.2020	74°56.9′ S	164°06.8′ E	15	53.3 333 3
2	12.09.2020	74°48.2′ S	166°00.4′ E	106	
3	12.11.2020	74°34.0′ S	171°00.1′ E	123	15.4
4	13.12.2020	77°22.4′ S	176°17.9′ E	155	1.3
5	14.12.2020	75°26.4′ S	176°17.9′ E	150	30.7
6	14.12.2020	74°31.5′ S	179°11.0′ W	103	1.0
7	15.12.2020	76°40.0′ S	179°11.9′ W	7	14.3
8	16.12.2020	77°41.6′ S	179°00.3′ W	28	21.4

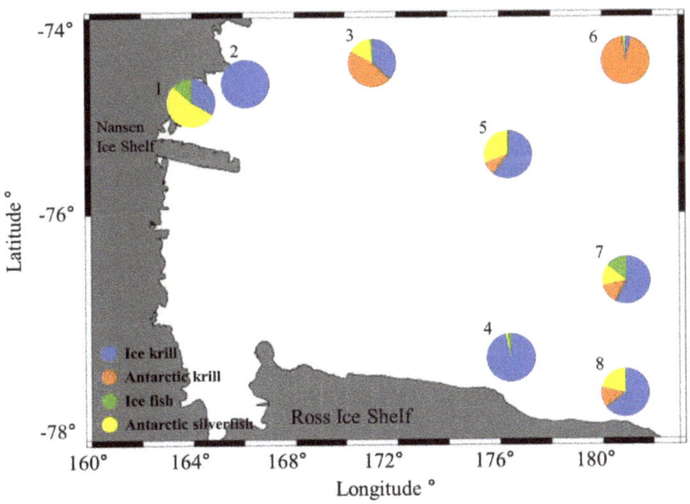

Figure 6. Distribution of caught species according to trawl station.

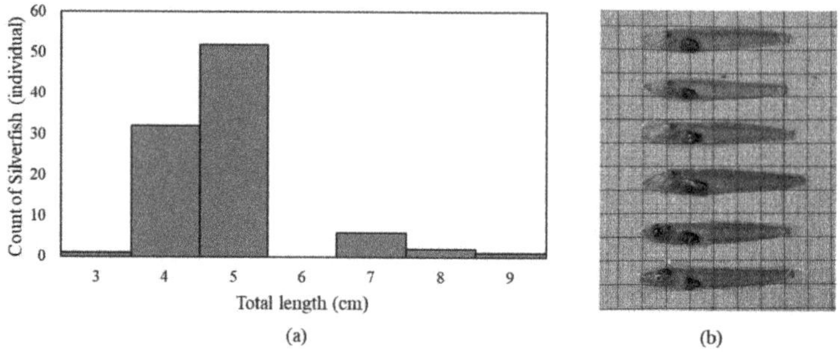

Figure 7. Antarctic silverfish sampled in December 2020 using frame midwater trawl. (**a**) Total length distribution. (**b**) Photograph of sampled Antarctic silverfish.

The maximum TS of Antarctic silverfish, calculated from the theoretical model of hydroacoustic scattering in line with their variation in posture angle [−45° to +45°], refers to the highest value between such posture angles of −45° to 45°; the average TS is the value calculated using PDF for a posture angle of −5 ± 15° [16]. The relationship between the total length and maximum TS of Antarctic silverfish was as follows:

$TS = 52.13\log_{10} TL - 119.00$ at 38 kHz,
$TS = 52.10\log_{10} TL - 112.67$ at 120 kHz
And $TS = 46.32\log_{10} TL - 107.24$ at 200 kHz

The relationship between the total length and average TS was as follows:

$TS = 47.74\log_{10} TL - 118.44$ at 38 kHz,
$TS = 47.65\log_{10} TL - 115.43$ at 120 kHz
And $TS = 41.37\log_{10} TL - 110.21$ at 200 kHz

The maximum TS of the reference TS, which was calculated with the frequency-specific TS proportional to the second power of length, was −92.93 dB at 38 kHz, −86.63 dB at 120 kHz, and −85.89 dB at 200 kHz; the average TS of Antarctic silverfish was −100.00 dB at 38 kHz, −93.00 dB at 120 kHz, and −106.90 dB at 200 kHz.

The vertical distribution of the NASC values for Antarctic silverfish in the nearshore waters of the Antarctic Ross Sea is shown in Figure 8. The detection range at a frequency of 200 kHz was considered in 10 m intervals from a depth of 15 m to a depth of 155 m. The vertical distribution in February 2018 showed low NASC values in all water layers. The vertical distribution in December 2018 showed the highest value of 4.29 m^2/nm^2 at a depth of 25 m. The NASC values in March 2020 were highest at the surface with 8.3 m^2/nm^2, but NASC values lowered with increasing depth. The values in December 2020 were highest at the surface, and higher values were found at depths of 50 and 60 m. Most high NASC values were found within 20 m. The vertical distribution of NASC values of the Antarctic silverfish in the polynya area in the Antarctic Ross Sea was as follows: strong NASC values were found at the surface in December 2018; in March 2020, NASC values tended to increase with depth. NASC values in December 2020 were found to be high at a depth of 20 m and higher at depths of 50 and 80 m. Both nearshore and polynya waters showed a strong distribution at the surface.

The horizontal distribution of the fish is shown in Figures 8 and 9, where the color of the circle indicates the size of the NASC (Figures 9 and 10). In February 2018, NASC values in the Ross Sea were generally low in intensity. The NASC values for juvenile Antarctic silverfish in the Ross Sea in December 2018 were stronger near the sea ice, and the strongest intensity was found near −75° latitude and 172° longitude. The NASC values for juvenile Antarctic silverfish in the Ross Sea in March 2020 were generally low, similar to the values in December 2018. The horizontal distribution of juvenile Antarctic silverfish in the Ross Sea in December 2020 was mostly even, and a strong intensity was found between −74° and −75° latitude. The distribution of juvenile Antarctic silverfish was stronger in December compared with February and March. The horizontal distribution of juvenile Antarctic silverfish in the polynya waters in December 2018 was found to be mostly low, with higher intensity in the north of the survey area. In March 2020, the horizontal distribution of juvenile Antarctic silverfish in the polynya waters was largely in the north of the Drygalski Trough, located at −75° latitude and 163–164° longitude. In December 2020, the horizontal distribution of juvenile Antarctic silverfish in the polynya waters showed a strong intensity near −78° latitude and 168–176° longitude, while the average NASC of the fish species per vessel was generally even.

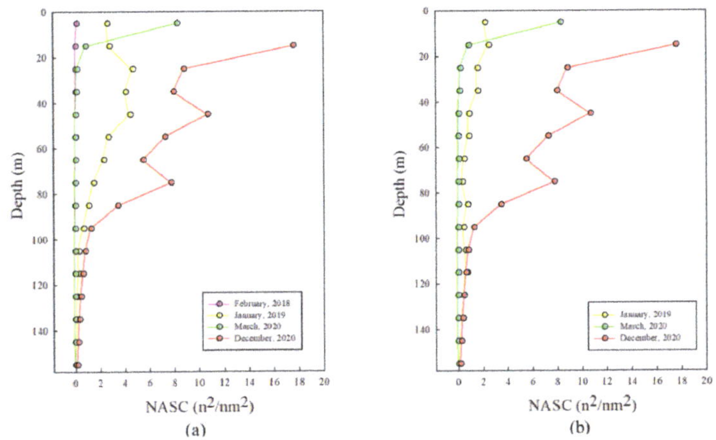

Figure 8. Vertical distribution of Antarctic silverfish NASC values. (**a**) Ross Sea and (**b**) polynya waters.

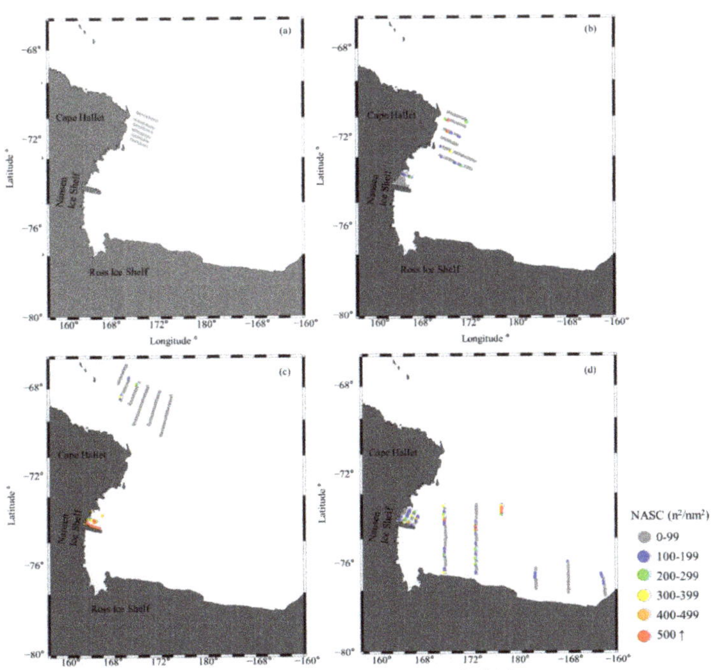

Figure 9. Horizontal distribution of Antarctic silverfish NASC values in the Ross Sea. (**a**) February 2018, (**b**) December 2018, (**c**) March 2020, and (**d**) December 2020.

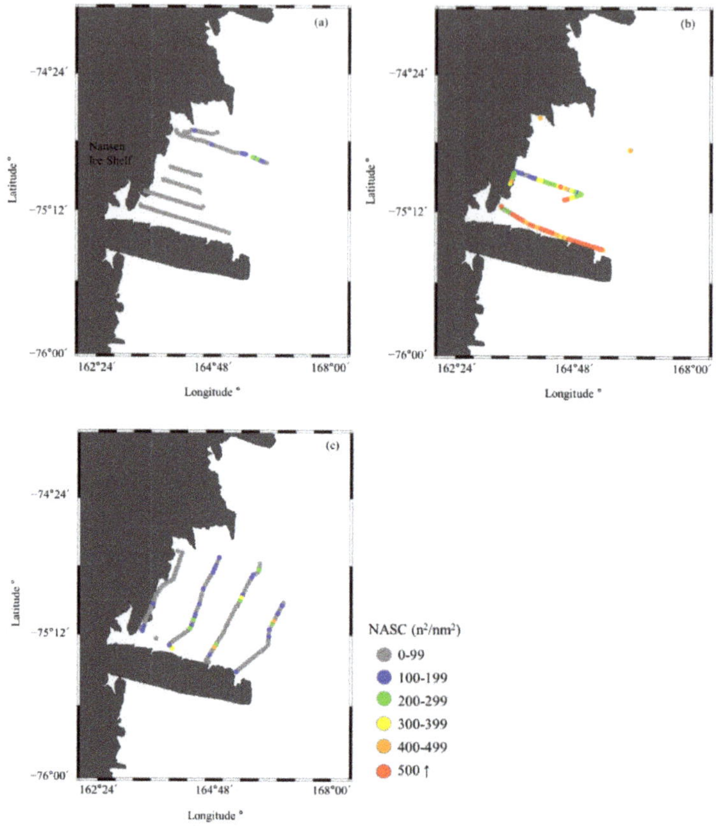

Figure 10. Horizontal distribution of Antarctic silverfish NASC values in polynya waters. (**a**) December 2018, (**b**) March 2020, and (**c**) December 2020.

In February 2018, the average density of the fish species in the nearshore waters of the Antarctic Ross Sea was 0.004 (g/m^2). In December 2018, the average density of the fish species was 0.058 (g/m^2) in the nearshore waters of the Antarctic Ross Sea and 0.008 (g/m^2) in the polynya waters. In March 2020, the average density of the fish species in the Antarctic Ross Sea was 0.029 (g/m^2) in the nearshore waters and 1.468 (g/m^2) in the polynya waters, confirming that the density was 50 times higher in the polynya waters than in the nearshore waters. In December 2020, the average density of the fish species in the Antarctic Ross Sea was 0.269 (g/m^2) in the nearshore waters and 0.337 (g/m^2) in the polynya waters. Higher densities were found in the polynya waters compared with nearshore waters, except in December 2018.

4. Discussion

Based on the same length, juvenile Antarctic silverfish have similar acoustic backscatter strength to Antarctic krill and show similar primary distribution depths to ice krill [19]. The current study examined post-larval Antarctic silverfish of 3–10 cm in length. The lengths of the collected krill were substantially different from those of Antarctic silverfish, and the acoustic scattering strengths were therefore also assumed to differ substantially (Table 6).

Table 6. Length distribution of Antarctic krill, ice krill, and Antarctic silverfish, sampled in December 2022.

	Antarctic Krill (*Euphausia superba*)	Ice Krill (*Euphausia crystallorophias*)	Antarctic Silverfish (*Pleuragramma antarcticum*)
Minimum length (cm)	3.14	1.77	3.42
Maximum length (cm)	4.95	3.92	9.40
Average length (cm)	4.10	2.79	4.83
Standard deviation	3.24	4.51	0.96

By comparing the MVBS (200–120 kHz) difference in the range 3–5 cm, where the lengths overlapped, the difference of MVBS (200–120 kHz) between Antarctic silverfish and Antarctic krill (*E. superba*) became similar as the lengths increased (Figure 11). Therefore, future studies should be conducted on the acoustic backscatter strength of Antarctic silverfish, Antarctic krill, and ice krill to develop algorithms that can more clearly distinguish the three species. Additionally, krill are characterized by large-scale clustering, whereas juvenile Antarctic silverfish tend to be more dispersed [20,21]. Therefore, studies of their clustering or dispersion patterns may also be helpful in distinguishing them.

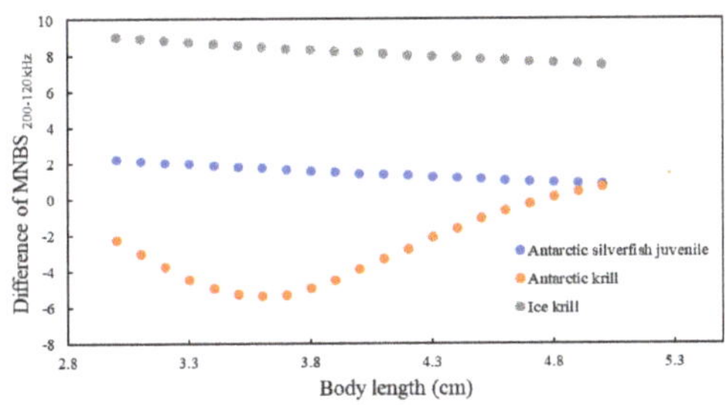

Figure 11. Comparison of the MVBS (200–120 kHz) values in the range of 3–5 cm among Antarctic silverfish, Antarctic krill, and ice krill.

The fishing gear used in the December 2018 survey was a frame trawl, which is designed to prevent the net from being damaged by vessel speed and currents. The length of Antarctic silverfish caught in this survey was 3–9 cm. Unlike previous studies focusing on post-larval and adult Antarctic silverfish, this study only targeted juvenile fish [22]. However, it was not easy to handle large sampling equipment, such as frame trawls, and move to target depths in Antarctic waters [23]. In the future, frame trawls may need to be equipped with real-time monitoring depth sensors to determine depth movements.

In a previous study by La Mesa et al. [3] that targeted Antarctic silverfish and estimated acoustic scattering intensities, the fish were divided into adult fish and larvae to estimate the acoustic scattering features; however, as the sample size employed in the current study corresponds to the size classified as fish larvae in the previous study, only acoustic scattering features of fish larvae were compared. In previous studies, the body shape of Antarctic silverfish is described as a spherical shape based on body depth, and the current study extended this into a cylindrical shape for body depth and width. The ratio of body width to depth of the Antarctic silverfish was approximately 0.08; since body width is relatively small, we assumed that there would be a large difference in acoustic scattering intensities. A previous study and the present study showed that acoustic scattering intensities increased in line with higher frequencies, which is a feature of fish species without a swimbladder [21].

O'Driscoll et al. [21] indicated that adult Antarctic silverfish are mainly distributed at depths of 100–400 m, whereas juvenile fish are mainly distributed in shallow waters at a depth of 80 m. Since echo signals were identified by applying the acoustic backscatter strength of juvenile Antarctic silverfish in the current study, we assumed that the echo signals did not appear in deeper waters.

Vacchi et al. [10] showed that Antarctic silverfish spawn near sea ice. In the present study, a large number of fertilized eggs of Antarctic silverfish were found near Terra Nova, the survey area. Antarctic silverfish were found to be distributed near sea ice. The phenomenon is speculated to be due to the fact that this research was conducted during the Antarctic summer when Antarctic silverfish is attracted by the influx of phytoplankton caused by warmer temperatures and melting sea ice.

Polynya is a geographical term indicating an ice-free area in polar oceans where sea ice is present and remains ice-free almost all year round. In this area, the biomass growth and primary production of phytoplankton increase due to the restrictions from physical phenomena such as wind, glaciers, and heat. In the survey area, the Terra Nova Bay polynya and the Ross Ice Shelf polynya are formed, which may explain the high distribution of Antarctic silverfish near polynya. Furthermore, both the Ross Ice Shelf polynya and Terra Nova Bay polynya are reported to produce the maximum amount of sea ice in March [24], which may explain the high fish distribution near the polynya in March 2020. In other survey periods, such as February and December 2018, the sea ice area in the Antarctic Ross Sea was the least developed of the corresponding months from 1981 to 2010 [25]. In particular, the sea ice rarely developed in February compared with December, which may explain the low density of Antarctic silverfish due to less development in its spawning ground.

5. Conclusions

In this study, the vertical and horizontal distribution of juvenile Antarctic silverfish was investigated using the MVBS method. The study area encompassed regions where Antarctic silverfish coexist with Antarctic krill and ice krill. The frequency range applied was based on the estimated acoustic backscattering strength of Antarctic silverfish using the KRM. The analysis of the acoustic backscattering strength of juvenile Antarctic silverfish using the KRM model revealed that as the frequency increased, the TS also increased. For species like Antarctic silverfish, which lack swimbladders, the swimming angle is an essential variable influencing TS. However, research on the swimming angle of juvenile Antarctic silverfish is currently limited. Therefore, it is essential to investigate the swimming angle of Antarctic silverfish schools in the future. The spatiotemporal distribution results indicate that they predominantly inhabit the upper 80 m of the water column, with higher intensity observed near ice-covered areas. Antarctic silverfish is a species known to spawn near ice, suggesting a need for specific research on the relationship between Antarctic silverfish and sea ice in the future. This study estimated the TS of Antarctic silverfish, and it confirmed that the difference in MVBS between Antarctic krill and ice krill is distinct. Despite these differences, the echogram analysis still occasionally misidentified krill as Antarctic silverfish. However, the distribution patterns of Antarctic silverfish and krill were markedly different. Consistent with the findings of the previous study, this study similarly reveals that krill forms schools, while juvenile Antarctic silverfish exhibit a dispersed distribution pattern [21]. Although an analyst can manually exclude krill based on these patterns, problems arise due to notable variations in results depending on the analyst. Therefore, it is necessary to develop algorithms for the precise determination of Antarctic silverfish distribution patterns in the future. We believe this study can offer foundational data on the abundance of juvenile Antarctic silverfish.

Author Contributions: Conceptualization, S.L. and K.L.; methodology, S.L.; software, W.O.; validation, H.S.L., J.-H.K. and W.S.; formal analysis, S.L.; investigation, W.O., H.S.L. and W.S.; resources, K.L.; data curation, S.L. and W.O.; writing—original draft preparation, S.L.; writing—review and editing, S.L.; visualization, S.L.; supervision, K.L.; project administration, K.L.; funding acquisition, H.S.L. and J.-H.K. All authors have read and agreed to the published version of the manuscript.

Funding: This research was supported by the Korea Institute of Marine Science & Technology Promotion (KIMST) funded by the Ministry of Oceans and Fisheries (KIMST-20220547).

Institutional Review Board Statement: Ethics Committee: CNU laboratory animal research center. Approval code: CNU IACUC-YS-2021-8. Approval date: 28 October 2021.

Data Availability Statement: No new data were created or analyzed in this study. Data sharing is not applicable to this article.

Acknowledgments: We acknowledge the support and dedication of the captain and crew of IBRV ARAON for completing the field work with positive energy. We would like to thank Euna Yoon and Hyungbeen Lee for providing valuable advice on the analysis of the KRM model. Finally, we are grateful to Sunyoung Oh and Geunchang Park for providing valuable advice, counsel, and guidance.

Conflicts of Interest: The authors declare no conflicts of interest.

References

1. Corso, A.D.; Steinberg, D.K.; Stammerjohn, S.E.; Hilton, E.J. Climate drives long-term change in Antarctic Silverfish along the western Antarctic Peninsula. *Commun. Biol.* **2022**, *5*, 104. [CrossRef]
2. La Mesa, M.; Eastma, J.T.; Vacchi, M. The role of notothenioid fish in the food web of the Ross Sea shelf waters: A review. *Polar Biol.* **2004**, *27*, 321–338. [CrossRef]
3. La Mesa, M.; Catalano, B.; Russo, A.; Greco, S.; Vacchi, M.; Azzali, M. Influence of environmental conditions on spatial distribution and abundance of early life stages of Antarctic silverfish, *Pleuragramma antarcticum* (Nototheniidae), in the Ross Sea. *Antarct. Sci.* **2010**, *22*, 243–254. [CrossRef]
4. Sutton, C.P.; Horn, P.L. A preliminary assessment of age and growth of Antarctic silverfish. *CCAMLR Sci.* **2011**, *18*, 75–86.
5. Tavernier, E.; Giraldo, C. Trophic ecology of early developmental stages of Antarctic silverfish. In *The Antarctic Silverfish: A Keystone Species in a Changing Ecosystem*; Springer: Berlin/Heidelberg, Germany, 2017; pp. 113–130.
6. Gon, O.; Heemstra, P.C. *Fishes of the Southern Ocean*; JBL Smith Institute of Ichthyology: Grahamstown, South Africa, 1990; pp. 313–316. [CrossRef]
7. Giraldo, C.; Cherel, Y.; Vallet, C.; Mayzaud, P.; Tavernier, E.; Moteki, M.; Koubbi, P. Ontogenic changes in the feeding ecology of the early life stages of the Antarctic silverfish (*Pleuragramma antarcticum*) documented by stable isotopes and diet analysis in the Dumont d'Urville Sea (East Antarctica). *Polar Sci.* **2011**, *5*, 252–263. [CrossRef]
8. Hubold, G. Stomach contents of the Antarctic silverfish *Pleuragramma antarcticum* from the southern and eastern Weddell Sea (Antarctica). *Polar Biol.* **1985**, *5*, 43–48. [CrossRef]
9. Koubbi, P.; O'Brien, C.; Loots, C. Spatial distribution and inter-annual variations in the size frequency distribution and abundances of *Pleuragramma antarcticum* larvae in the Dumont d'Urville Sea from 2004 to 2010. *Polar Sci.* **2011**, *5*, 225–238. [CrossRef]
10. Vacchi, M.; La Mesa, M.; Dalu, M.; MacDonald, J. Early life stages in the life cycle of Antarctic silverfish, *Pleuragramma antarcticum* in Terra Nova Bay, Ross Sea. *Antarct. Sci.* **2004**, *16*, 299–305. [CrossRef]
11. Park, G.C.; Oh, W.S.; Oh, S.Y.; Lee, K.H. Acoustic scattering characteristics of chub mackerel (*Scomber japonicus*) by KRM model. *J. Korean Soc. Fish. Ocean Technol.* **2022**, *58*, 32–38. [CrossRef]
12. Do, M.A.; Surti, A.M. Estimation of dorsal aspect target strength of deep-water fish using a simple model of swimbladder backscattering. *J. Acoust. Soc. Am.* **1990**, *87*, 1588–1596. [CrossRef]
13. Vacchi, M.; Bottaro, M.; DeVries, A.L. A nursery area for the Antarctic silverfish *Pleuragramma antarcticum* at Terra Nova Bay (Ross Sea): First estimate of distribution and abundance of eggs and larvae under the seasonal sea-ice. *Polar Biol.* **2012**, *35*, 1573–1585. [CrossRef]
14. Yoon, S.T. Responses of the Ross Sea to the climate change: Importance of observations in the Ross Sea, Antarctica. *Ocean Polar Res.* **2022**, *44*, 69–82. [CrossRef]
15. Foote, K.G.; Traynor, J.J. Comparison of walleye pollock target strength estimates determined from in situ measurements and calculations based on swimbladder form. *J. Acoust. Soc. Am.* **1988**, *83*, 9–17. [CrossRef]
16. Hwang, K.S.; Lee, K.H.; Hwang, B.K. Verification and application of Target Strength for Japanese anchovy (*Engraulis japonicas*) by theoretical acoustic scattering model. *J. Korean Soc. Fish. Ocean Technol.* **2012**, *48*, 487–494. [CrossRef]
17. Azzali, M.; Leonori, I.; Biagiotti, I.; De Felice, A.; Angiolillo, M.; Bottaro, M.; Vacchi, M. Target strength studies on Antarctic silverfish (*Pleuragramma antarcticum*) in the Ross Sea. *CCAMLR Sci.* **2010**, *17*, 75–104.
18. Ryan, T.E.; Downie, R.A.; Kloser, R.J.; Keith, G. Reducing bias due to noise and attenuation in open-ocean echo integration data. *ICES J. Mar. Sci.* **2015**, *72*, 2482–2493. [CrossRef]
19. La, H.S.; Lee, H.; Kang, D.; Lee, S.; Shin, H.C. Ex situ echo sounder target strengths of ice krill *Euphausia crystallorophias*. *Chin. J. Oceanol. Limnol.* **2015**, *33*, 802–808. [CrossRef]
20. Tarling, G.A.; Klevjer, T.; Fielding, S.; Watkins, J.; Atkinson, A.; Murphy, E.; Leaper, R. Variability and predictability of Antarctic krill swarm structure. *Deep Sea Res. Part I Oceanogr. Res. Pap.* **2009**, *56*, 1994–2012. [CrossRef]

21. O'Driscoll, R.L.; Macaulay, G.J.; Gauthier, S.; Pinkerton, M.; Hanchet, S. Distribution, abundance and acoustic properties of Antarctic silverfish (*Pleuragramma antarcticum*) in the Ross Sea. *Deep Sea Res. Part II Top. Stud. Oceanogr.* **2011**, *58*, 181–195. [CrossRef]
22. Granata, A.; Cubeta, A.; Guglielmo, L.; Sidoti, O.; Greco, S.; Vacchi, M.; La Mesa, M. Ichthyoplankton abundance and distribution in the Ross Sea during 1987–1996. *Polar Biol.* **2002**, *25*, 187–202. [CrossRef]
23. Oozeki, Y.; Hu, F.; Kubota, H.; Sugisaki, H.; Kimura, R. Newly designed quantitative frame trawl for sampling larval and juvenile pelagic fish. *Fish. Sci.* **2004**, *70*, 223–232. [CrossRef]
24. Tamura, T.; Ohshima, K.I.; Fraser, A.D.; Williams, G.D. Sea ice production variability in Antarctic coastal polynyas. *J. Geophys. Res. Ocean.* **2016**, *121*, 2967–2979. [CrossRef]
25. Copernicus. Sea Ice. Available online: https://climate.copernicus.eu/sea-ice (accessed on 6 May 2023).

fishes

MDPI

Article

Vocalization Pattern and Echolocation Signal Characteristics of Yangtze Finless Porpoise (*Neophocaena asiaeorientalis asiaeorientalis*) in Captivity

Jia Chen [1,†], Haiying Liang [1,†], Danqing Lin [2], Jialu Zhang [2], Dong Li [2], Kun Ye [2], Wenfei Lu [2] and Kai Liu [1,2,*]

1 National Demonstration Center for Experimental Fisheries Science Education, Shanghai Ocean University, Shanghai 201306, China
2 Key Laboratory of Freshwater Fisheries and Germplasm Resources Utilization, Ministry of Agriculture and Rural Affairs, Freshwater Fisheries Research Center, Chinese Academy of Fishery Sciences, Wuxi 214081, China
* Correspondence: liuk@ffrc.cn
† These authors contributed equally to this work.

Abstract: The Yangtze finless porpoise (*Neophocaena asiaeorientalis asiaeorientalis*, YFP) possesses the ability to detect distance through echolocation signals, and its sonar signal signature is adjusted to detect different targets. In order to understand the vocal characteristics of YFPs in different behavioral states and their differential performance, we recorded the vocal activities of YFPs in captivity during free-swimming, feeding, and nighttime resting and quantified their signal characteristic parameters for statistical analysis and comparison. The results showed that the number of vocalizations of the YFPs in the daytime free-swimming state was lower than that in the feeding and nighttime resting states, and the echolocation signals emitted in these three states showed significant differences in the −10 dB duration, −3 dB bandwidth, −10 dB bandwidth, and root-mean-square (RMS) bandwidth. Analysis of the resolution of the echolocation signals of the YFPs using the ambiguity function indicated that their distance resolution could reach the millimeter level. These results indicate that the echolocation signal characteristics of YFPs present diurnal differences and that they can be adjusted with changes in their detection targets. The results of this study can provide certain scientific references and foundations for the studies of tooth whale behavioral acoustics, and provide relevant scientific guidance for the conservation and management of YFPs.

Keywords: Yangtze finless porpoise; echolocation signal; behavioral state; diurnal variation; ambiguity function

Key Contribution: In this study, we analyzed and compared the vocalization pattern and echolocation signal characteristics of the Yangtze finless porpoise (*Neophocaena asiaeorientalis asiaeorientalis*) in captivity during free-swimming, feeding, and nighttime resting, as well as the discriminative ability of their echolocation signals, to explore the differences in the changes in the echolocation signals of the YFPs in different behavioral states with a view to providing certain references for acoustic studies of the YFPs' biological behaviors.

1. Introduction

The Yangtze finless porpoise (*Neophocaena asiaeorientalis asiaeorientalis*, YFP) is a small toothed whale endemic to China. It is distributed in the middle and lower reaches of the Yangtze River, as well as Dongting Lake and Poyang Lake. It serves as an important indicator species for the health of the freshwater ecosystem of the Yangtze River and the status of its biodiversity [1]. YFPs were listed as critically endangered (CR) by the Species Survival Commission of the World Conservation Union (IUCN/SSC) in 2013 [2]. With only 1249 of their species remaining in the wild by 2022, their conservation remains important [3].

YFPs have evolved an effective echolocation system to adapt to the freshwater living environment, which mainly allows them to carry out activities necessary for survival such as localization and navigation, inter-individual communication and exchange, prey hunting, and escaping from enemies [4–6]. However, a number of human water-related activities generate tremendous noise [7,8], which affect YFPs' sonar activities to a certain extent and even threaten their survival activities. There is an urgent need to study the echolocation behavior of this group of toothed whales, which are frequently affected by human activities and are in endangered conditions.

In recent years, it has been demonstrated that harbor porpoises (*Phocoena phocoena*) use different echolocation strategies and biosonar parameters in two different environments for solving an otherwise identical target approach task, thus highlighting that biosonar adjustments are both range- and context-dependent [9]. Fang et al. demonstrated that the echolocation signal parameters of the YFPs differed among the three environments: a captive tank, a netted pen, and field water of the Yangtze River, indicating that the YFPs adapted to their echolocation signals depending on their surroundings [10]. Acoustic monitoring of YFPs in some field waters of the Yangtze River revealed that their nighttime sonar activities were greater than those of daytime, which may be related to their nocturnal foraging [11,12]. It has also been shown that the sonar activity of YFPs may be related to the changes in their surrounding environments [13]. However, little is known about whether YFPs adjust their sonar signal activity in response to behavioral activities or diurnal variations.

In addition, one study used differently spaced objects to allow harbor porpoises to discriminate between detections, and they were found to be able to resolve and discriminate closely spaced targets, suggesting a clutter rejection zone much shorter than their auditory integration time and that such clutter rejection is greatly aided by spatial filtering with their directional biosonar beam [14]. Ambiguity functions for processing radar signals have also been used to study the discriminatory ability of echolocation signals in toothed whales [15]. Studies on the ability to discriminate sonar signals have mainly focused on toothed whales living in the ocean, while studies on the YFPs in this area have not yet been reported.

Studying the echolocation signal activity and characteristics of YFPs in different behavioral states is crucial for understanding their survival strategies. Due to the large spatial extent of the field waters, the turbidity of the water, and the small body size of the YFPs, the implementation of acoustic monitoring and behavioral observations had limitations that were not conducive to the conduct of this study. Therefore, we studied the sonar activity patterns and signal characteristics of the YFPs in different behavioral states in a relatively controlled environment in captivity. This will help to understand the vocalization patterns and changing characteristics of the YFPs and provide a reference basis for the relevant departments to formulate targeted management policies and protection measures for the YFPs based on the patterns and characteristics of their sonar activities.

2. Materials and Methods

2.1. Subjects

The experimental research subjects were two male YFPs in the captive breeding base, and the data collection period was from March to April 2023.

2.2. Experimental Procedure

Data acquisition equipment: ① A fixed underwater high-frequency acoustic event recorder (i.e., A-tag, ML200-AS8 Marine Micro Technology, Saitama, Japan) was used for the monitoring of the YFPs vocalization events. Only specific high-frequency acoustic signal events were recorded during operation and were detected at a distance of approximately 300 m [16]. The sensitivity of the hydrophone is close to the main frequency of YFPs vocalization at 120 kHz [17]; electronic bandpass filtering is 55–235 kHz to eliminate underwater noise signals outside the frequency band of YFPs sonar signals; The sampling rate was set to 2 kHz, i.e., the pulse signal was recorded every 0.5 ms [18].

② The underwater eco-acoustic recorder (SoundTrap 300HF, Ocean Instruments Ltd., Auckland, New Zealand) was used to collect the YFPs echolocation signals, with the Sample Rate set to 576 kHz, the analog-to-digital converter (ADC) at 16 bits, and the Preamp gain set to high.

These two types of underwater sound recorders have been widely used in passive acoustic studies of YFPs [19–21]. During data collection, one SoundTrap 300HF and one A-tag were placed 1.5 m from the surface of the water in one corner of the rearing tank. Acoustic signal acquisition in this study was carried out uninterruptedly during the observation period, and the accompanying law enforcement recorder was used for recording; audible event acquisition continued uninterruptedly for 24 h. The hydrophones did not affect the daily activities of the YFPs, ensuring that the animals were free to move around and eliminating human interference as much as possible.

Nine days were randomly selected for the study of temporal changes in the number of vocalization events, and the 24 h were roughly divided into three categories based on the state of the YFPs appearing in one hour (e.g., Table 1). Divers would clean and vacuum the rearing pool from 8:00 to 9:00 daily, which was excluded due to the presence of anthropogenic disturbance. The feeding baits were all live *Hemiculter leucisculus*, and toy balls were provided daily in the captive tank.

Table 1. Time period (hours) of the YFPs activity in three states.

	Time Period
Free-swimming State	6:00–7:00; 7:00–8:00; 9:00–10:00; 11:00–12:00; 13:00–14:00; 15:00–16:00; 17:00–18:00
Feeding State	10:00–11:00; 12:00–13:00; 14:00–15:00; 6:00–17:00; 19:00–20:00; 21:00–22:00; 0:00–1:00
Nighttime Resting State	1:00–2:00; 2:00–3:00; 3:00–4:00; 4:00–5:00; 5:00–6:00; 18:00–19:00; 20:00–21:00; 22:00–23:00; 23:00–24:00

The echolocation signals of the YFPs were collected under three experimental conditions: the first was the free-swimming condition (i.e., when no one was interfering in the daytime, the two YFPs were in a variety of swimming postures, mostly swimming counterclockwise along the edge of the feeding pool in pairs with uniform swimming speeds.); the second was the feeding condition (i.e., when the feeders were feeding them live fish during the daytime, the two YFPs exhibited more complex behavior, with alternating feeding periods). (We picked up the signals emitted during its detection of live fish.) and the third was the nighttime resting condition (i.e., when no one was interfering at night, the behavior of the two YFPs was mostly accompanied by swimming, and their swimming speed was uniform and slowed down.).

2.3. Data Acquisition and Analytical Processing

① Data analysis of vocal events: The acoustic data collected with the A-tag were exported and converted into a format using the Logger tools (v5.03) software, and the acoustic signal events of the YFPs were extracted from the background noise events using a customized program based on the Igor Pro 7 (WaveMetrics, Lake Oswego, OR, USA) software. Usually, the echolocation signals of YFPs are expressed in the form of pulse sequences, and one pulse sequence emitted will contain five to hundreds of pulses [22], and sound pressure levels vary uniformly, and the inter-click intervals (ICIs) vary between 20 and 70 ms, whereas the changes in the sound pressure levels and ICIs of the underwater noise are irregular, so that the recorded acoustic signal pulse characteristics can be used to determine the vocalization events of YFPs [23]; In addition, the durations of the YFPs pulse sequences are generally less than 130 ms [24]. We screened the YFPs signals according to the above rule.

② Echolocation signal characterization: The data collected using SoundTrap were decompressed into .wav format files and imported into its accompanying analysis software, dBwav 1.3, and the data were calibrated according to the end-to-end sensitivity provided on the official website (http://www.oceaninstruments.co.nz/, accessed on 6 August 2023).

According to the high energy distribution of echolocation signals in YFPs, the corresponding audio clips were intercepted against the spectrogram, and the information was exported.

In order to select the echolocation signal as closely as possible to the propagation axis, we referred to some criteria set in previous studies [17,25,26]. The main steps to further selecting and analyzing the dBwav 1.3 intercepted audio clips include: (1) As far as possible, ensure that the audio was selected to be recorded with the animal's head facing the hydrophone; (2) Echolocation signal sequence sequences were independent signal sequences, excluding the superposition of multiple signal sequences; (3) The single pulse signal with the highest amplitude and typical characteristics of a toothed whale's waveform was selected, and the part with a clear sinusoidal waveform and smooth envelope structure was intercepted, while the rest of the reverberating structure was eliminated; (4) The intercepted echolocation signals with signal-to-noise ratio (SNR) higher than 5 dB were calculated using a MatLab R2022a customized code for further analysis [27].

Using the above methods, the author selected 300 single-pulse echolocation signals for the YFPs in the three different behavioral states: free-swimming, feeding, and nighttime resting, for a total of 900 echolocation signals. The peak frequency (which is the frequency that corresponds to the energy maximum of the signal spectrum), the −3 dB bandwidth (−3 dB_BW, which is the frequency range bandwidth at 3 dB less than the peak power), and the −10 dB bandwidth (−10 dB_BW, which is the frequency range bandwidth at 10 dB less than the peak power) of the intercepted echolocation signals were manually processed using Praat6211; the centroid frequency (which is the frequency at which the power spectrum is divided into two equal parts), the root-mean-square bandwidth (RMS_BW, which is the spectral standard deviation of the centroid frequency), and −10 dB duration (which is the duration between two points that are −10 dB below the peak of the signal waveform) of the signals were processed using custom MatLab code [28,29]. The bandwidth of the signal can be parameterized using −3 dB_BW and −10 dB_BW as well as RMS_BW [6,28,29]. The parameters of the spectrum setting in Praat were as follows: the visible frequency range was 0~288 kHz, the analysis mode was Fourier, the window function type was Hanning, and the window length was 0.0002 s.

The acoustic parameters of the signals in each state were statistically analyzed using SigmaPlot 14.0, and significant differences were analyzed using one-way ANOVA and the Mann–Whitney U-test. The results are expressed as mean ± standard deviation, and the data were plotted using SigmaPlot 14.0. For the description of the characteristic distributions of echolocation signal parameters, reference was made to the descriptions of the histogram distributions in previous studies of YFPs [10,17].

Referring to Li et al. [30] and Chen et al. [31], the echolocation signals' discriminative ability of YFPs was analyzed by custom MatLab code using the ambiguity function. The function describes the time-frequency characteristics of the echolocation signal. It is commonly used for the characterization of sonar signals and is expressed as

$$|\chi(\tau,\xi)| = \left| \int_{-\infty}^{\infty} s(t)s^*(t+\tau)e^{-j2\pi\xi t}dt \right|$$

where τ is the signal time delay and ξ is the signal Doppler shift.

The results of the ambiguity function's characterization of the signal are mainly in the form of a 3D ambiguity plot expressing the degree of ambiguity of the neighboring targets and its 2D cross-section at the peak drop to 70.7%, which is called the ambiguity plot. The two jointly reflect the signal's resolving power for distance (time delay value) and velocity (Doppler shift) of neighboring targets and the detection accuracy of the distance and velocity quantities. The time delay resolving power is the time difference $\Delta\tau = 2\tau_0$ between two intersections of the ambiguity plot and the τ-axis, while the frequency shift resolving power is the frequency shift difference $\Delta\xi = 2\xi_0$ between two intersections of the ambiguity map and the ξ-axis. The detection accuracy is then half of the resolving power (i.e., τ_0 and ξ_0). Therefore, the steeper the 3D ambiguity function plot surface and the smaller the cross-sectional area of the ambiguity plot, the easier it is to discriminate between neighboring targets [32].

3. Results

3.1. Comparison of Vocalization Number Characteristics of the YFPs in Three Behavioral States

Screening of the YFPs vocalization events over a 9-day period and counting the percentage of each hour's vocalization to the 24 h total showed that the YFPs vocalization activity changed at different hours of the day. The number of vocalization events of the YFPs in a 24 h period showed significant peaks in five periods: 12:00–13:00, 16:00–17:00, 19:00–20:00, 21:00–22:00, and 0:00–1:00, indicating strong vocalization activities (see Figure 1). Excluding the condition of special human interference (suction by divers from 8:00 to 9:00 a.m.), the three states show that the average proportion of vocalizations per hour in the 24 h total was lower during the daytime free swimming period (3.60%) than during feeding (5.09%) and the nighttime resting period (4.20%), and was highest during the feeding state (see Figure 2).

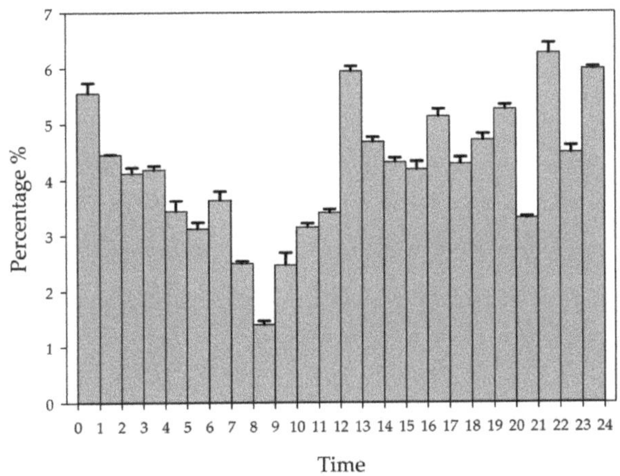

Figure 1. Temporal characteristics of the intensity of vocalization events of the YFPs.

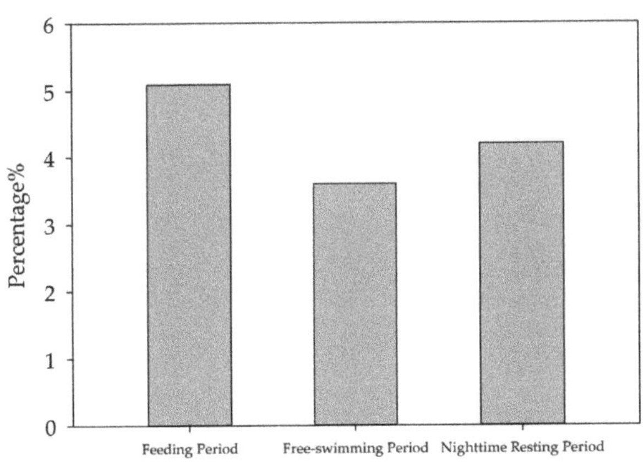

Figure 2. Intensity characteristics of vocalization events in three states of the YFPs.

3.2. Comparison of Echolocation Signal Parameters of the YFPs under Three Behavioral States

The acoustic parameters such as peak frequency, centroid frequency, signal bandwidth, and −10 dB duration of the echolocation signals under the three states were quantified and counted, and the parameters of the three groups were compared and analyzed using one-way ANOVA and then analyzed using the Mann–Whitney U-test for the differences among the groups. The acoustic parameters of the echolocation signals of the YFPs showed variability in the signal bandwidths (−3 dB_BW, −10 dB_BW, and RMS_BW) and the −10 dB duration under the three behavioral states, and the results of their one-way ANOVA are shown in Table 2. The mean, standard deviation, and variability of the echolocation signal parameters for each activity state are listed in Table 3, and the distribution is shown in Figure 3. The peak frequency of the echolocation signals was predominantly distributed at 125–135 kHz in all three states of free-swimming, feeding, and nighttime resting, accounting for 67%, 65%, and 66.67%, respectively. In the distribution of the −10 dB duration, two peaks of 30–40 μs and 50–60 μs were observed in the signal distribution during both the free-swimming and nighttime resting conditions, which accounted for 21%, 19.67%, 24.33%, and 23%, respectively, while the echolocation signals of the YFPs during feeding were mainly distributed in the range of 60–70 μs. In the distribution of −3 dB_BW, a peak of 15–25 kHz was observed in the signal distribution during the nighttime resting condition, while a peak of 10–20 kHz was observed in the signal distribution during the free-swimming and feeding states, which was 57.66% in the free-swimming state and 65.33% in the feeding state. The RMS_BW of the echolocation signals in the three states of free-swimming, feeding, and nighttime resting were mainly distributed in the range of 10–20 kHz, which accounted for 32%, 44%, and 38.33%, respectively. The distribution of the echolocation signals during feeding was higher in the number of low bandwidth values compared to the other two states and lower in the number of low −10 dB durations than the other two states.

Table 2. One Way ANOVA results of acoustic parameters of echolocation signals of the YFPs in three states.

	p Value
Peak frequency (kHz)	0.171
−3 dB_BW (kHz)	<0.001 ***
−10 dB_BW (kHz)	<0.001 ***
−10 dB duration (μs)	<0.001 ***
Centroid frequency (kHz)	0.059
RMS_BW (kHz)	0.017 *

* $p < 0.05$ *** $p < 0.001$.

Table 3. Calculated echolocation signal characteristics of the YFPs in three states.

	State (I)	State (J)	Mean ± S.D. (I)	Mean ± S.D. (J)	p Value
Peak frequency (kHz)	Free-swimming	Feeding	130.59 ± 5.25	129.77 ± 5.66	0.064
	Free-swimming	Nighttime Resting	130.59 ± 5.25	130.05 ± 5.54	0.249
	Feeding	Nighttime Resting	129.77 ± 5.66	130.05 ± 5.54	0.540
−3 dB_BW (kHz)	Free-swimming	Feeding	19.78 ± 5.60	19.14 ± 5.57	0.011 *
	Free-swimming	Nighttime Resting	19.78 ± 5.60	22.04 ± 6.02	<0.001 ***
	Feeding	Nighttime Resting	19.14 ± 5.57	22.04 ± 6.02	<0.001 ***
−10 dB_BW (kHz)	Free-swimming	Feeding	34.57 ± 8.30	33.79 ± 7.90	0.204
	Free-swimming	Nighttime Resting	34.57 ± 8.30	37.80 ± 10.12	<0.001 ***
	Feeding	Nighttime Resting	33.79 ± 7.90	37.80 ± 10.12	<0.001 ***
−10 dB duration (μs)	Free-swimming	Feeding	54.52 ± 18.88	59.20 ± 18.70	0.001 **
	Free-swimming	Nighttime Resting	54.52 ± 18.88	49.11 ± 15.94	<0.001 ***
	Feeding	Nighttime Resting	59.20 ± 18.70	49.11 ± 15.94	<0.001 ***
Centroid frequency (kHz)	Free-swimming	Feeding	122.56 ± 10.75	124.14 ± 10.12	0.094
	Free-swimming	Nighttime Resting	122.56 ± 10.75	124.24 ± 9.64	0.132
	Feeding	Nighttime Resting	124.14 ± 10.12	124.24 ± 9.64	0.820
RMS_BW (kHz)	Free-swimming	Feeding	28.90 ± 15.35	26.14 ± 14.00	0.035 *
	Free-swimming	Nighttime Resting	28.90 ± 15.35	25.89 ± 13.42	0.037 *
	Feeding	Nighttime Resting	26.14 ± 14.00	25.89 ± 13.42	0.920

* $p < 0.05$ ** $p < 0.01$ *** $p < 0.001$.

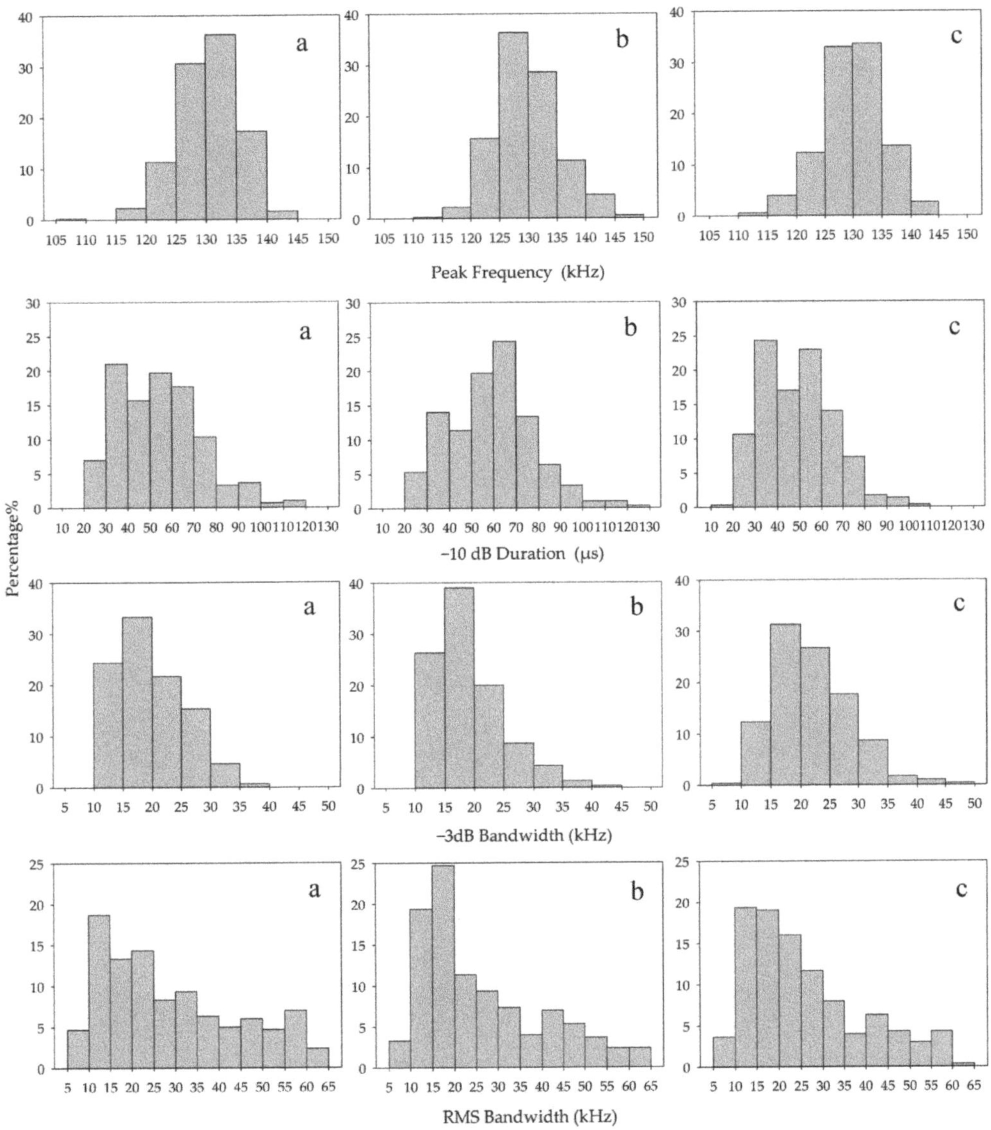

Figure 3. Histograms of echolocation signal characteristics of the YFPs (Free-swimming (a); Feeding (b); Nighttime Resting (c)).

3.3. Echolocation Signal Resolution of the YFPs

To describe the discriminative ability of YFP echolocation signals, a typical sequence of the YFPs echolocation signals (see Figure 4a) was selected, for which Figure 4b shows the time frequency characteristic map. The single-pulse signal with the largest amplitude (see Figure 4c) was intercepted and analyzed. The 3D ambiguity function plot and ambiguity plot of the echolocation signal of the YFPs are shown in Figure 5, and the peak in the 3D ambiguity plot is obvious. From the actual data of the ambiguity plot (see Figure 5b), it can be seen that the time-delay measurement accuracy τ_0 of the YFP sonar signal was about 0.569 μs, the distance resolution ability of the signal to the target was $\rho = c \cdot \tau_0$, and the speed of sound in the Yangtze River was 1465.93 m·s^{-1} (water temperature of 15 °C, water depth of 5 m); thus, ρ = 0.834 mm. Notably, the distance resolution was still quite high. The time-delay accuracy measurements of the YFP echolocation signal sequence (see Figure 4a) show that the time-delay accuracy of the individual echolocation signals varied between 0.5 and 0.9 µs (see Figure 6). Figure 7 shows the 3D ambiguity function plots and ambiguity plots of typical echolocation signals for the three behavioral states, and the results show that they reached the millimeter level in all three states.

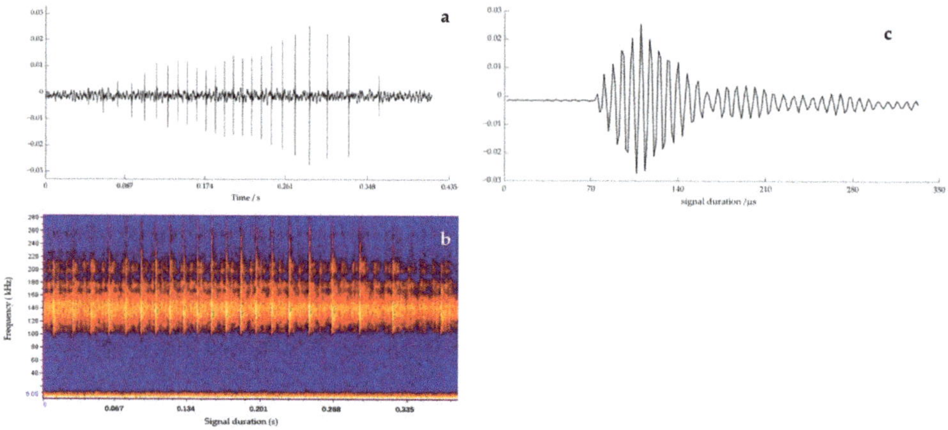

Figure 4. Characteristics of the echolocation signal sequence of the YFP (Waveform diagram (**a**); Time frequency characteristic (**b**); Single pulse waveform (**c**)).

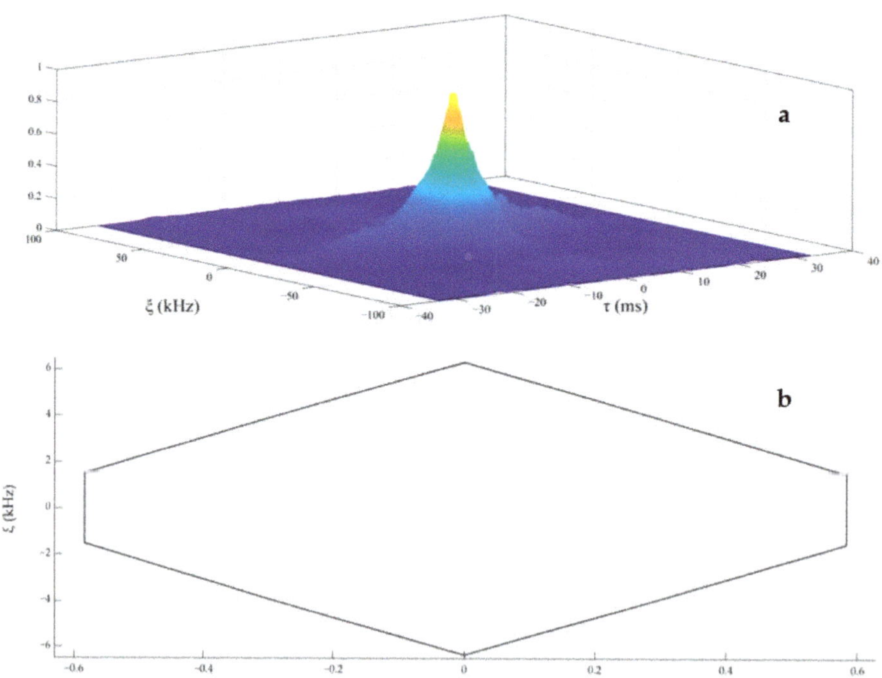

Figure 5. The 3D ambiguity function plot (**a**) and ambiguity plot (**b**) of a single echolocation signal of the YFP.

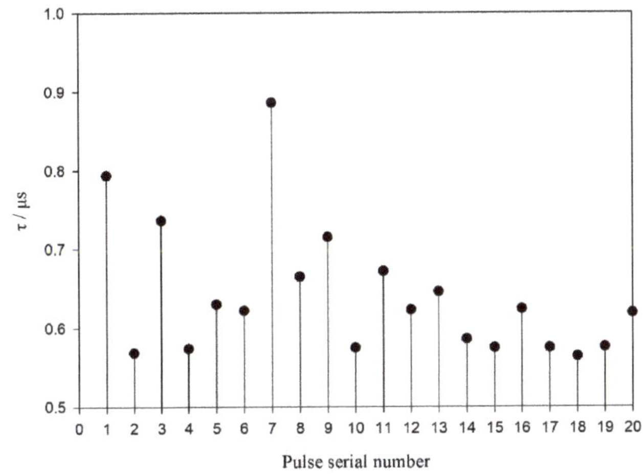

Figure 6. Variation in time delay measurement accuracy of the echolocation signal sequence of the YFP.

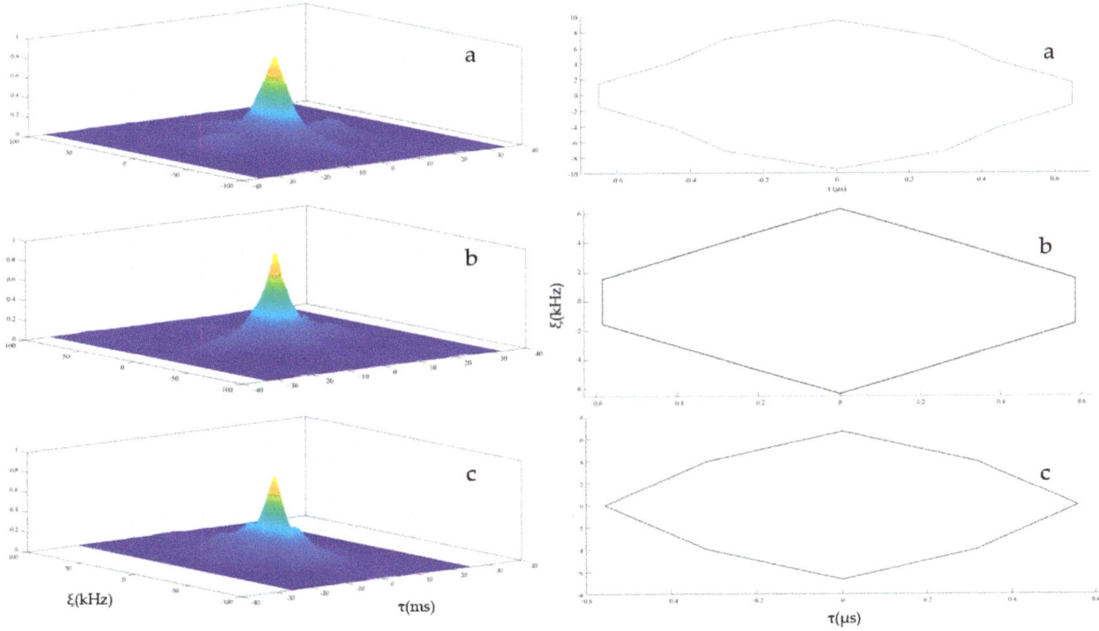

Figure 7. The 3D ambiguity function plots and ambiguity plots of individual echolocation signals of the YFPs (Free-swimming (a), Feeding (b), Nighttime Resting (c)).

4. Discussion

4.1. Analysis of the Diurnal Variation Pattern of the Number of Vocalization Events of the YFPs

In this study, the 9-day monitoring results of the vocalization events of the YFPs showed that there were diurnal variations in their vocal activity, with five peak time periods (12:00–13:00, 16:00–17:00, 19:00–20:00, 21:00–22:00, and 0:00–1:00) corresponding to the five feeding periods. The YFPs emitted echolocation signals significantly more frequently during feeding than in the free-swimming condition, a phenomenon that has also been observed in studies of toothed whales such as beluga whales (*Delphinapterus leucas*) [33], and bottlenose dolphins (*Tursiops truncatus*) [34]. This may be related to the fact that YFPs emit predatory signals during feeding, especially buzz signals when catching prey [20], leading to a significant increase in pulse signals; it is also possible that the presence of live bait during feeding caused the YFPs to increase the frequency of emitting echolocation signals. In their study on the factors influencing the frequency of vocalizations of YFPs, Serres et al. found that the YFPs emitted the highest number of echolocation signals under the condition of the presence of live fish [13].

The results of this experiment showed that the average number of vocalization events per hour during nighttime resting accounted for a higher percentage than during the daytime in the free-swimming condition and that four periods of higher vocalization existed at night, mainly at 18:00–19:00, 23:00–24:00, 1:00–2:00, and 3:00–4:00. In studies of YFPs in the field, it has also been found that more vocalization activities of YFPs were monitored at night [8,11,12,20], and it is believed that such phenomena are related to the tendency of YFPs to prefer nocturnal foraging activities. However, the nighttime feeding period was eliminated from this study, no food was provided, and the animals spent most of their time in an undisturbed resting state. Regarding the reasons for the high percentage of average hourly vocalizations during nighttime resting, the authors believe that first, the animal may have been in the clear water body of the captive tank and, so, some visual aids were present for observation, which could reduce vocal activities. Second, it is possible that the vocal behaviors were increased prior to the nighttime resting for the purpose of communicating with their companions, promoting the synchronization of slow swimming and detection of the surrounding environment. Cetacean sleep is characterized by single-hemisphere slow-wave sleep, and the sleep cycle is around 1 h, which explains the results in the present study of nocturnal resting with periods of higher vocalizations and their intervals of around 1–2 h [35]. A similar phenomenon has been seen in studies of bottlenose dolphin vocalizations, where bottlenose dolphins increase vocal interactions before resting to ensure synchrony in slow swimming [36,37].

The YFP is a highly social animal that requires vocal behavior in order to perform functions such as socialization, orientation, and the transmission of information. The number of vocal events of YFPs changes under different behavioral states, which reflects their physiological and psychological states in different contexts [13,38]. The frequency of vocalizations of YFPs shows seasonal variations due to breeding activities, which may be attributed to their need to use their voices to attract the opposite sex and establish pair bonding [13,39]. Toothed whales also vocalize differently to provide feedback in the face of negative external noise threat stimuli [13]. In this study, the YFPs increased their vocal behavior during their feeding and nocturnal resting periods. This provides a number of references for the formulation of conservation policies. For example, strategies such as controlling the speed of vessels or even temporarily restricting travel are implemented in the waters where porpoises frequently feed in the wild, as well as during preferred foraging hours or night resting hours, in order to reduce the impacts of anthropogenic water-related activities. However, only a sample size of nine days from March to April was selected for the study, not covering the four seasons of the year. Therefore, selecting a sample size with a larger seasonal span and a larger number of days to carry out further studies will help to comprehensively understand the sonar activity patterns of the YFPs and facilitate scientific management in a captive environment.

4.2. Analysis of Echolocation Signals Characterization of the YFPs in Three Behavioral States

The YFPs' echolocation signals showed significant differences in the −10 dB duration, −3 dB_BW, −10 dB_BW, and RMS_BW values during the free-swimming, feeding, and nighttime resting conditions. The bandwidth averages (−3 dB_BW, −10 dB_BW, and RMS_BW) for the echolocation signals of the YFPs in the feeding state were lower than those in the free-swimming state, while the −10 dB duration averages of the signals were higher than the other two states. The author suggests that the reason for these differences may be the inconsistency in the detection targets required for the YFPs in the three behavioral states. The live bait targets that need to be detected in the feeding state are more finely tuned and variable in location than the surroundings that need to be detected during free swimming and nighttime resting. Studies have shown that the characteristics of the echolocation signals of toothed whales can be adjusted according to their detection target range. A reduction in the signal bandwidth facilitates the animal to extract the signal from the noise and improve the SNR, while an increase in the signal duration enhances the energy of the detected signal without increasing the signal amplitude [40]. The YFPs actively emit echolocation signals with low bandwidth and high duration during feeding to adjust the SNR and energy of the signals to improve their detection accuracy.

The echolocation signal characteristics of the YFPs in this study showed differences between the nighttime resting and the free-swimming state during the daytime, with an overall high bandwidth and low mean duration of their nighttime signals, which reduced the energy of the detected signal [40]. In their research on melon-headed whales (*Peponocephala electra*), foreign scholars have found that their echolocation signals were adjusted according to day and night, possibly to adapt to changes in ambient noise or to enhance the detection of targets [41]. However, no investigation of diurnal ambient background noise was conducted in this study in order to confirm whether this was related to their adaptation to changes. The author suggests that the differences in the signal characteristics of this type of YFPs may have arisen as an adaptation to the behavioral pattern of resting and socializing at night.

In their research on the vocalization and behavior of toothed whales, Wei et al. studied the characteristics of communication signals (low-frequency signals) of bottlenose dolphins in a free-swimming state and a training state and found that there was a significant change in the proportion of communication signal categories (low-frequency signals) of bottlenose dolphins in the two behavioral states; furthermore, there was a variability in the acoustic parameters of the signals, indicating that there was an inevitable connection between their vocalization and behavior [42]. We provided further evidence that toothed whale vocalizations are tuned to behavioral traits in terms of high-frequency signals. However, in this study, only one single hydrophone was used for acoustic signal acquisition, and despite the author's rigorous signal selection, it was still impossible to avoid recording signals that deviate from the propagation axis, which may bias the results to some extent. Thus, further studies of signals from porpoises in different behavioral states using a range of receivers (e.g., hydrophone arrays) will help to increase the understanding of the acoustics of porpoise behavior and provide more insight into the use of sonar signals.

4.3. Analysis of Echolocation Signal Resolution of the YFPs

In the study of YFPs echolocation system for target detection, scholars have found that YFPs perform echolocation vocalizations once every 5.1 s on average, and their detection distance was projected based on the vocalization interval, which showed that 90% of the acoustic detection distances are less than 77 m [16]. However, the resolving power of the echolocation signals of YFPs has not been investigated. We analyzed their resolving power and detection accuracy in terms of the time domain, frequency domain, time frequency analysis, and ambiguity function.

In this study, the echolocation signals of the YFPs in all three behavioral states indicated strong anti-reverberation ability and high distance resolution. Compared to the results of Li et al. on bottlenose dolphins, beluga whales, and sperm whales (*Pseudorca crassidens*) in captivity, the distance discrimination of YFPs is at the same level as that of these three toothed whales [30]. Chen et al. conducted a study on the bionic bottlenose dolphin echolocation signal, and the results showed that the length of the bottlenose dolphin echolocation signal in the natural environment is shorter, which improves its speed discrimination. In an artificial breeding environment, for the bottlenose dolphin echolocation signal, the peak frequency was lower, the bandwidth was narrower, and the side flap of the 3D ambiguity function was smaller, which is beneficial for anti-reverberation and distance discrimination [31]. The YFPs constantly changed the characteristics of each pulsed signal during signaling vocalizations, as in the case of bottlenose dolphins and sperm whales [30]. Qing studied and analyzed the echolocation signal sequence strings of beluga whales and bottlenose dolphins, and the results proved that they may achieve target-optimized detection and identification through transmitting pulse strings with different types of energy distribution [43]. This also indicates that YFPs can achieve optimal detection and identification of a target by continuously changing the echolocation signal characteristics during target detection.

YFPs are often found in shallow water depths and are therefore subject to a great deal of unwanted echo interference. However, porpoises can still become entangled or drown in nets that their biosonar is capable of detecting [44], and we understand that the YFPs are also threatened in this way. The distance discrimination of the YFPs in this study demonstrated that their echolocation signals were able to discriminate nets, which is consistent with the results of other toothed whale net detection experiments [45]. When a task is difficult and attention-demanding, foraging performance can be constrained, and the detection of threats may be hindered [46]. The YFPs are similarly less alert to threatening objects, such as nets, when performing focused detection tasks. Although a ten-year fishing ban has been fully implemented in the Yangtze River since 2021, the remaining pollution, such as discarded fishing nets, still exists. It is extremely important to manage the ecological environment of the YFPs' frequented waters and cultivate the public's awareness.

5. Conclusions

In this study, we analyzed the vocalization patterns and signal characteristics of YFPs during free swimming, feeding, and nighttime resting, and gained a more in-depth understanding of their habits, such as active time, resting time, and feeding habits, which will enhance the accuracy of the field acoustic monitoring results and contribute to the monitoring and protection of the wild populations. This will provide relevant scientific guidance for the development of protection measures for YFPs, which will better manage anthropogenic water-related activities and effectively protect YFPs from threats. As a next step, the behavioral acoustic characteristics of YFPs in different habitats should be studied in depth over a longer time span using hydrophone arrays to investigate the relationship between their vocalizations and behavioral status in different habitats, as well as the effects of diurnal variations in the seasons on their sonar activities.

Author Contributions: Conceptualization, D.L. (Danqing Lin) and K.L.; Data curation, J.C., H.L. and J.Z.; Data analysis, J.C., W.L. and H.L.; Funding acquisition, D.L. (Danqing Lin) and K.L.; Investigation, J.C.; Methodology, J.C. and D.L. (Danqing Lin); Project administration, K.L.; Resources, D.L. (Danqing Lin) and K.L.; Software, J.C., K.Y. and D.L. (Dong Li); Supervision, K.L.; Validation, D.L. (Danqing Lin), D.L. (Dong Li) and K.L.; Writing—original draft, J.C.; Writing—review and editing, D.L. (Danqing Lin) and K.L. All authors have read and agreed to the published version of the manuscript.

Funding: This project was financed by the Central Public-interest Scientific Institution Basal Research Fund, CAFS (NO. 2023TD11); Protection of the Yangtze Finless Porpoise; and Publicizing Science (17230149).

Institutional Review Board Statement: We strictly followed the national regulations and policies for animal protection in China. The medical examinations and related experiments conducted were approved by competent authorities of the Department of Resource Environmental Protection, Yangtze River Basin Fisheries Regulation and Management Office, Ministry of Agriculture. Which reviewed and approved the procedures for animal observation and passive acoustic monitoring (2017 [185] and 2018 [183]).

Informed Consent Statement: Not applicable.

Data Availability Statement: Data are contained within the article.

Conflicts of Interest: The authors declare no conflicts of interest.

References

1. Zhou, X.; Guang, X.; Sun, D.; Xu, S.; Yang, G. Population genomics of finless porpoises reveal an incipient cetacean species adapted to freshwater. *Nat. Commun.* **2018**, *9*, 1276. [CrossRef]
2. Wang, D.; Turvey, S.; Zhao, X.; Mei, Z. *Neophocaena asiaeorientalis ssp. asiaeorientalis. The IUCN Red List of Threatened Species*; Version 3.1; IUCN: Gland, Switzerland; Cambridge, UK, 2013.
3. Yan, F.J.; Yu, W.J. The Yangtze finless porpoise population has rebounded to 1249. *Xinhua Daily Telegraph*, 1 March 2023; p. 006.
4. Wang, K.X.; Wang, D. Characteristics and Functions of Sound of the Yangtze Finless Porpolse (*Neophocaena phocaenoides*) in Captivity. *Tech. Acoust.* **1999**, *18*, 4.
5. Wang, D. A preliminary study on sound and acoustic behavior of the Yangtze Finless Porpoise, *Neophocaena Phocaenoides*. *Acta Hydrobiol. Sin.* **1996**, *20*, 127–133.
6. Busnel, R. (Ed.) *Animal Sonar Systems*; Springer Science & Business Media: Berlin/Heidelberg, Germany, 2013; Volume 28.
7. Zhang, T.C.; Ju, T.; Li, S.H.; Xie, Y.; Wang, D.; Wang, Z.T.; Wang, K.X. Navigation noise property of large vessels in Hechangzhou region of the Yangtze River and its potential effects on the Yangtze finless porpoise. *Acta Theriol. Sin.* **2018**, *38*, 543–550.
8. Duan, P.X.; Wang, Z.T.; Akamatsu, T.; Tregenza, N.; Li, G.Y.; Wang, K.X.; Wang, D. Anthropogenic activity, hydrological regime, and light level jointly influence temporal patterns in biosonar activity of the Yangtze finless porpoise at the junction of the Yangtze River and Poyang Lake, China. *Zool. Res.* **2023**, *44*, 919–931. [CrossRef]
9. Michael, L.; Peter, T.M. Context-dependent biosonar adjustments during active target approaches in echolocating harbour porpoises. *J. Exp. Biol.* **2019**, *222*, jeb206169.
10. Fang, L.; Wang, D.; Li, Y.; Cheng, Z.; Pine, M.K.; Wang, K.X.; Li, S.H. The source parameters of echolocation clicks from captive and free-ranging Yangtze finless porpoises (*Neophocaena asiaeorientalis asiaeorientalis*). *PLoS ONE* **2015**, *10*, e0129143. [CrossRef]
11. Wang, Z.; Akamatsu, T.; Wang, K.; Wang, D. The diel rhythms of biosonar behavior in the Yangtze Finless Porpoise (*Neophocaena asiaeorientalis asiaeorientalis*) in the port of the Yangtze River: The correlation between prey availability and boat traffic. *PLoS ONE* **2014**, *9*, e97907. [CrossRef]
12. Chen, M.; Yu, D.; Wang, K.; Zhang, K.; Wang, Z. Seasonal and diel activities of the Yangtze finless porpoise in natural and highly disturbed habitats: Implications for conservation planning of freshwater cetaceans. *Aquat. Conserv. Mar. Freshw. Ecosyst.* **2022**, *32*, 605–616. [CrossRef]
13. Serres, A.; Xu, C.; Hao, Y.G.; Wang, D. The click production of captive Yangtze Finless Porpoises (*Neophocaena asiaeorientalis asiaorientalis*) is influenced by social and environmental factors. *Animals* **2021**, *11*, 511. [CrossRef]
14. Chloe, E.M.; Laia, R.-D.; Peter, T.M. Directional biosonar beams allow echolocating harbour porpoises to actively discriminate and intercept closely-spaced targets. *J. Exp. Biol.* **2021**, *224*, jeb242779.
15. Thorpe, C.W.; Bates, R.H.; Dawson, S.M. Intrinsic echolocation capability of Hector's dolphin, *Cephalorhynchus hectori*. *J. Acoust. Soc. Am.* **1991**, *90*, 2931–2934. [CrossRef]
16. Akamatsu, T.; Wang, D.; Wang, K.; Li, S.; Dong, S.; Zhao, X.; Barlow, J.; Stewart, B.S.; Richlen, M. Estimation of the detection probability for Yangtze finless porpoises (*Neophocaena phocaenoides asiaeorientalis*) with a passive acoustic method. *J. Acoust. Soc. Am.* **2008**, *123*, 4403–4411. [CrossRef]
17. Li, S.H.; Wang, K.X.; Wang, D.; Akamatsu, T. Echolocation signals of the free-ranging Yangtze finless porpoise (*Neophocaena phocaenoides asiaeorientialis*). *J. Acoust. Soc. Am.* **2005**, *117*, 3288–3296. [CrossRef]
18. Li, S.H.; Wang, D.; Wang, K.X.; Taylor, E.A.; Cros, E.; Shi, W.J.; Wang, Z.T.; Fang, L.; Chen, Y.F.; Kong, F.M. Evoked-potential audiogram of an Indo-Pacific humpback dolphin (*Sousa chinensis*). *J. Exp. Biol.* **2012**, *215 Pt 17*, 3055–3063. [CrossRef]
19. Li, S.; Akamatsu, T.; Dong, L.; Wang, K.; Wang, D.; Kimura, S. Widespread passive acoustic detection of Yangtze finless porpoise using miniature stereo acoustic data-loggers: A review. *J. Acoust. Soc. Am.* **2010**, *128*, 1476–1482. [CrossRef]
20. Wang, Z.; Akamatsu, T.; Mei, Z.; Dong, L.; Imaizumi, T.; Wang, K.; Wang, D. Frequent and prolonged nocturnal occupation of port areas by Yangtze finless porpoises (*Neophocaena asiaeorientalis*): Forced choice for feeding? *Integr. Zool.* **2015**, *10*, 122–132. [CrossRef]
21. Zhang, K.; Zhang, P.; Xia, D.J.; Li, Z.H.; Wu, Z.; Yu, D.P.; Chen, M.M. Study on the utilization pattern of island-type habitat patches by Yangtze finless porpoise and analysis of potential factors. *Acta Hydrobiol. Sin.* **2023**, *48*, 1–10.
22. Sasaki-Yamamoto, Y.; Akamatsu, T.; Ura, T.; Sugimatsu, H.; Kojima, J.; Bahl, R.; Behera, S.; Kohshima, S. Diel changes in the movement patterns of Ganges River dolphins monitored using stationed stereo acoustic data loggers. *Mar. Mammal Sci.* **2013**, *29*, 589–605. [CrossRef]
23. Akamatsu, T.; Wang, D.; Wang, K.X.; Naito, Y. Biosonar behaviour of free-ranging porpoises. *Proc. R. Soc. B* **2005**, *272*, 797–801. [CrossRef]
24. Akamatsu, T.; Teilmann, J.; Miller, L.A.; Tougaard, J.; Dietz, R.; Wang, D.; Siebert, U.; Naito, Y. Comparison of echolocation behaviour between coastal and riverine porpoises. *Deep. Sea Res. Part II Top. Stud. Oceanogr.* **2007**, *54*, 290–297. [CrossRef]
25. Au, W.W.; Würsig, B. Echolocation signals of dusky dolphins (*Lagenorhynchus obscurus*) in Kaikoura, New Zealand. *J. Acoust. Soc. Am.* **2004**, *115*, 2307–2313. [CrossRef]
26. Fang, L.; Li, S.; Wang, K.; Wang, Z.; Shi, W.; Wang, D. Echolocation signals of free-ranging Indo-Pacific humpback dolphins (*Sousa chinensis*) in Sanniang Bay, China. *J. Acoust. Soc. Am.* **2015**, *138*, 1346–1352. [CrossRef]
27. Wang, Y.Y.; Wang, X.Y.; XU, X.H.; Song, Z.Z.; Yang, W.Y.; Zhang, Y. Analysis of echolocation from Indo-Pacific humpback dolphin (*Sousa chinensis*) and finless porpoise (*Neophocaena asiaeorientalis sunmeri*). *Acta Acust.* **2021**, *46*, 423–430.

28. Madsen, P.T.; Carder, D.A.; Bedholm, K.; Ridgway, S.H. Porpoise clicks from a sperm whale nose—Convergent evolution of 130 kHz pulses in toothed whale sonars? *Bioacoustics* **2005**, *15*, 195–206. [CrossRef]
29. Madsen, P.T.; Wahlberg, M. Recording and quantification of ultrasonic echolocation clicks from free-ranging toothed whales. *Deep. Sea Res. Part I Oceanogr. Res. Pap.* **2007**, *54*, 1421–1444. [CrossRef]
30. Li, H.; Gao, D.Z.; Lin, J.H.; Chi, J. Analyses of click signal characteristics of captive bottlenose dolphins, beluga whales and false killer whales. *Tech. Acoust.* **2019**, *38*, 24–31.
31. Chen, S.; Niu, F.Q.; Lin, C.L.; Liu, W. Click signal of bottlenose dolphin and its simulation analysis under different environments. *Tech. Acoust.* **2019**, *38*, 7.
32. Tian, T. *Sonar Technology*, 2nd ed.; Harbin Engineering University Press: Harbin, China, 2010; pp. 22–30.
33. Fish, M.P.; Mowbray, W.H. Production of underwater sound by the white whale or beluga, *Delphinapterus leucas (Pallas)*. *J. Mar. Res.* **1962**, *20*, 149–162.
34. Niu, F.Q.; Yang, Y.M.; Wen, H.T.; Liu, Z.W. Vocalization and signal characteristic of bottlenose dolphin. *Tech. Acoust.* **2011**, *30*, 148–152.
35. Lyamin, O.I.; Manger, P.R.; Ridgway, S.H.; Mukhametov, L.M.; Siegel, J.M. Cetacean sleep: An unusual form of mammalian sleep. *Neurosci. Biobehav. Rev.* **2008**, *32*, 1451–1484. [CrossRef]
36. Kremers, D.; Jaramillo, M.B.; Böye, M.; Lemasson, A.; Hausberger, M. Nocturnal vocal activity in captive bottlenose dolphins (*Tursiops truncatus*): Could dolphins have presleep choruses? *Anim. Behav. Cogn.* **2014**, *1*, 464–469. [CrossRef]
37. Therrien, S.C.; Thomas, J.A.; Therrien, R.E.; Stacey, R. Time of Day and Social Change Affect Underwater Sound Production by Bottlenose Dolphins (*Tursiops truncatus*) at the Brookfield Zoo. *Aquat. Mamm.* **2012**, *38*, 65–75. [CrossRef]
38. Serres, A.; Hao, Y.G.; Wang, D. Swimming features in captive odontocetes: Indicative of animals' emotional state? *Behav. Process.* **2020**, *170*, 103998. [CrossRef]
39. Wei, Z.; Wang, D.; Zhang, X.F.; Wang, K.X.; Chen, D.Q.; Zhao, Q.Z.; Wang, X. Observation on some sexual behavior of the Yangtze finless porpoise (*Neophocaena phocaenoides asiaerientals*) in captivity. *Acta Theriol. Sin.* **2004**, *24*, 98–102.
40. Morisaka, T.; Karczmarski, L.; Akamatsu, T.; Sakai, M.; Dawson, S.; Thornton, M. Echolocation signals of Heaviside's dolphins (*Cephalorhynchus heavisidii*). *J. Acoust. Soc. Am.* **2011**, *129*, 449–457. [CrossRef]
41. Baumann-Pickering, S.; Roch, M.A.; Wiggins, S.M.; Schnitzler, H.U.; Hildebrand, J.A. Acoustic behavior of melon-headed whales varies on a diel cycle. *Behav. Ecol. Sociobiol.* **2015**, *69*, 1553–1563. [CrossRef]
42. Wei, C.; Xu, X.M.; Zhang, Y.; Niu, F.Q. Comparison study on whistle characteristics of captive bottlenose dolphins (*Tursiops turncatus*) during swimming and under training. *Acta Acust.* **2014**, *39*, 452–458. (In Chinese)
43. Qing, X. *Research on the Mechanisms of Biosonar Operation in the Representative Hunting Behaviour of Cetaceans*; Harbin Engineering University: Harbin, China, 2021.
44. Read, A.J.; Drinker, P.; Northridge, S. Bycatch of marine mammals in U.S. and global fisheries. *Conserv. Biol.* **2006**, *20*, 163–169. [CrossRef]
45. Kastelein, R.A.; Au, W.W.L.; de Haan, D. Detection distances of bottom-set gillnets by harbour porpoises (*Phocoena phocoena*) and bottlenose dolphins (*Tursiops truncatus*). *Mar. Environ. Res.* **2000**, *49*, 359–375. [CrossRef]
46. Dukas, R.; Kamil, A.C. The cost of limited attention in blue jays. *Behav. Ecol.* **2000**, *11*, 502–506. [CrossRef]

fishes

MDPI

Article

Investigations on Target Strength Estimation Methods: A Case Study of Chub Mackerel (*Scomber japonicus*) in the Northwest Pacific Ocean

Zhenhong Zhu [1,2,3], Jianfeng Tong [1,4,5,*], Minghua Xue [1], Chuhan Qiu [1], Shuo Lyu [1] and Bilin Liu [1,4,5]

1 College of Marine Living Resource Sciences and Management, Shanghai Ocean University, Shanghai 201306, China; zhuzhenhong1996@163.com (Z.Z.); d220200065@st.shou.edu.cn (M.X.); q15836082195@163.com (C.Q.); 13624518054@163.com (S.L.); bl-liu@shou.edu.cn (B.L.)
2 Wuxi Fisheries College, Nanjing Agricultural University, Wuxi 214081, China
3 Key Laboratory of Freshwater Fisheries and Germplasm Resources Utilization, Ministry of Agriculture and Rural Affairs, Freshwater Fisheries Research Center, Chinese Academy of Fishery Sciences, Wuxi 214081, China
4 National Engineering Research Center for Oceanic Fisheries, Shanghai 201306, China
5 Key Laboratory of Sustainable Exploitation of Oceanic Fisheries Resources, Ministry of Education, Shanghai 201306, China
* Correspondence: jftong@shou.edu.cn

Abstract: Target strength (TS) is an acoustic property of individual marine organisms and a critical factor in acoustic resource assessments. However, previous studies have primarily focused on measuring TS at narrowband, typical frequencies, which cannot meet the requirements of broadband acoustic technology research. Additionally, for marine fish, conducting in situ TS measurements is challenging due to environmental constraints. Rapidly freezing and preserving fish samples for transfer to the laboratory is a common method currently used. However, the impact of freezing preservation during transportation on the swimbladder morphology and TS of swimbladder-bearing fish remains unclear. This study investigated the differences in swimbladder morphology and TS of Chub mackerel (*Scomber japonicus*) before and after freezing. Then, we compared different TS measurement methods through ex situ TS measurements (45–90 kHz, 160–260 kHz) and the Kirchhoff-ray mode model (KRM) simulations (1–300 kHz) and studied the broadband scattering characteristics of Chub mackerel based on the KRM model. The results showed that the morphology of the swimbladder was reduced after freezing, with significant changes in swimbladder height and volume. However, the trends of TS were not consistent and the changes were small. The difference between the KRM model and ex situ measurements was -0.38 ± 1.84 dB, indicating good applicability of the KRM. Based on the KRM results, the TS exhibited significant directivity, with fluctuations gradually decreasing and stabilizing as frequency increased. In the broadband mode, the relationship between TS and body length (L) of Chub mackerel was TS = 20log(L) − 66.76 ($30 > L/\lambda > 10$). This study could provide a reference for acoustic resource estimation and species identification of Chub mackerel in the Northwest Pacific Ocean.

Keywords: Chub mackerel; swimbladder morphology; target strength; broadband scattering response; acoustic resource estimation

Key Contribution: The study examined the changes in the swimbladder morphology and TS before and after freezing, compared the differences between ex situ measurements and model estimates, and obtained the broadband scattering characteristics of Chub mackerel.

1. Introduction

The fisheries acoustic technology based on echosounders aims to conduct remote sensing detection of fishery resources and even marine ecosystems without biological and environmental damage, providing fishery-independent scientific data [1,2]. Compared to biological sampling methods, fisheries acoustic technology can provide vertical profile data for detailed analysis of water layers, offering much broader survey coverage [3]. One of the main focuses of fisheries acoustic research is converting acoustic data into biomass data to assess the abundance of target species in the survey area. The key parameter affecting the accuracy of biomass data is target strength (TS; dB re 1 m^2), which represents the ability of an individual organism to reflect sound waves [4]. Additionally, with the widespread use of broadband scientific echosounders that can transmit a wide and continuous frequency range, acoustic data resolution, and signal-to-noise ratio (SNR) have been increased, significantly improving the ability to target classification based on backscatter spectra [5,6]. The backscatter spectrum of a single individual, which is the variation of TS at the continuous frequency ($TS(f)$), correlates with the backscattering from schools ($Sv(f)$) of the same species: $Sv(f) = TS(f) + 10\log\rho$, where ρ is the biomass density in schools [7]. Therefore, to achieve species identification based on broadband acoustic technology, it is first necessary to study the spectral characteristics of individual target strength.

Fish TS is influenced by various factors, including fish species, acoustic frequency, presence or absence of a swimbladder, body length, and swimming posture [8–11]. For swimbladder-bearing fish, the density contrast between the swimbladder and the body tissue greatly impacts the acoustic pulse, with up to 95% of the backscatter attributable to the swimbladder [8]. The study on Argentine anchovy (*Engraulis anchoita*) TS by Madirolas et al. [12], based on the Prolate Spheroidal Model (PSM), indicated that the variation in fish TS at different incident angles was consistent with the TS variation of the swimbladder, while the TS of the fish body remained low. The study by Li et al. [13] on Thorn fish (*Terapon jarbua*) using the Kirchhoff-ray mode (KRM) also corroborated this perspective. Changes in swimbladder morphology can significantly influence TS and lead to considerable differences in biomass estimates [14,15].

Current research on TS estimation mainly includes direct measurement methods, such as in situ [16,17] and ex situ measurements [18–20], as well as acoustic scattering model estimation methods based on morphological parameters [21–23]. In situ measurements can obtain TS data in the natural state, but are limited by the field environment, interference from other organisms, and the inability to control the length distribution of the fish, making it challenging to accurately establish TS–length relationships [24]. Ex situ measurements are conducted by setting up experimental environments and using methods such as net cages and tethering to measure fish echo information. Although this approach allows for linking TS data with fish length, it alters the natural state of the fish. Using samples that have been anesthetized or frozen may lead to changes in their morphology, potentially causing deviations in TS measurements [25]. While direct measurement experiments can better reflect the TS under real conditions, both in situ and ex situ measurements present significant challenges in data collection, and sample size is also a crucial factor limiting the accuracy of the obtained TS data [26]. Acoustic scattering models approximate the geometric shape based on the morphological characteristics of the fish and estimate TS using scattering theory. This approach facilitates the study of TS variations under different influencing factors [9]. However, the accuracy of the models is limited by the precision of geometric measurements. Additionally, each model has its advantages and limitations, and the choice of model depends on the target shape and specific requirements [27]. Generally, as the accuracy of models continues to improve, research on TS based on acoustic scattering models has been increasing. However, the comparison between actual measurements and model estimates remains noteworthy. Hazen et al. [28] compared the ex situ TS measurements with the predictions of the backscatter model, examining the differences between the two methods in terms of fish length, tilt, depth, and frequency. Peña et al. [29] constructed an accurate fish body morphology based on MRI scan results and compared it with the standard TS obtained from ex situ experiments. Numerous studies have demonstrated the differences and applicability between various TS measurement methods.

In conducting research related to the TS of marine fish, the complexity of the environment and the interference from various factors affecting TS make in situ TS measurements challenging. Therefore, ex situ measurements in laboratory or acoustic scattering model estimations are often used instead [30,31]. Both of these methods require transferring biological samples from the survey area to the laboratory, often facing long-term preservation challenges. Additionally, for swimbladder-bearing fish, it is even more important to maintain the integrity of the swimbladder morphology. Different preservation methods may lead to structural changes in the fish body or compression of the swimbladder, resulting in variations in TS. Sobradillo et al. [15] placed the samples in liquid nitrogen and then stored them frozen at −15 °C, but obtaining sufficient liquid nitrogen on board was more difficult. Rapid freezing of fresh samples preserved in a seawater-filled sampling device is the current common method [32,33], but attention should be paid to the differences in fish swimbladder morphology before and after freezing. Currently, there is a lack of research on the impact of different sample preservation methods on swimbladder morphology.

The confluence of cold and warm currents in the Northwest Pacific Ocean creates abundant fishery resources, making it one of the highest-yielding marine fishing areas in the world [34]. As a significant pelagic fishery resource, Chub mackerel is widely distributed and possesses high economic value. It holds an important position among the various fishery resources in the Northwest Pacific Ocean and is under the jurisdiction of the North Pacific Fisheries Commission (NPFC) [35]. With the continuous development of global fisheries, the exploitation of Chub mackerel resources has been intensifying. According to the NPFC statistics, there has been a general declining trend in catches from the 1980s to the present, which has attracted closer attention from the NPFC and related researchers [36]. As early as 2015, the NPFC listed Chub mackerel as a priority species for management, and a working group was established to carry out research on the status of Chub mackerel stocks and conservation management. However, the assessment of Chub mackerel resources has not yet been completed [37]. The International Council for the Exploration of the Sea (ICES), the Commission for the Conservation of Antarctic Marine Living Resources (CCAMLR), and other organizations have established relevant acoustic research working groups. They utilize acoustic technology to provide references for the assessment and management of fisheries resources. However, the acoustic method has not been widely applied in the Northwest Pacific Ocean.

Target strength (TS) is a crucial aspect of fisheries acoustics research, directly affecting the accuracy of fishery resource assessments. The primary objectives of this study are: (a) to investigate the changes in swimbladder morphology and their impact on TS caused by freezing preservation of fish samples, by measuring the morphological parameters of the swimbladder and fish body before and after freezing; (b) to compare the differences between ex situ TS measurements and KRM model results, validating the applicability of the KRM model for TS measurement in Chub mackerel; and (c) to obtain the broadband scattering response characteristics and the relationship between TS and body length at typical frequencies for Chub mackerel. This study aims to provide fundamental data for the acoustic resource assessment of Chub mackerel and to support species identification research based on broadband acoustic technology, thereby promoting the application of fisheries acoustic technology in the management of fishery resources in the Northwest Pacific.

2. Materials and Methods

2.1. Biological Sample Collection

Fish samples were collected from a scientific survey of pelagic fishery resources in the Northwest Pacific Ocean (34–45° N, 147–165° E), carried by the scientific research vessel "Songhang" from 15 June to 30 July 2022. The study area was located at the confluence of the Kuroshio warm current and the Oyashio cold current, which was characterized by a rich aggregation of biological resources [38,39]. The survey deployed a four-panel fine mesh mid-water trawl, with primary dimensions of 434 × 97.1 m and a 40 mm stretch cod-end mesh, structured as a single bag. The deployment and position of the trawl, as well as the net opening size and other relevant details, were closely monitored and recorded using a trawl sounder [40]. Within the area, a total of 36 biological sampling stations were established. Each trawling operation lasted approximately 1.5 h, with an average depth of 44.98 ± 13.74 m. The vessel speed was approximately 4–5 knots.

At all sampling stations, Chub mackerel catches were randomly selected as active or undamaged and transferred to sampling bottles filled with seawater. A portion of the samples was randomly chosen to be dissected in the onboard biological laboratory, while the remainder was placed in the −20 °C freezer for rapid freezing and preservation, intended for research after the return voyage.

2.2. Morphological Measurements

The morphological parameters measured for each sample are shown in Figure 1, including fish body length (fbl), body width (fbw), body height (fbh), swimbladder length (sbl), swimbladder width (sbw), swimbladder height (sbh), and swimbladder tilt angle. The swimbladder tilt angle is defined as the angle between the central axes of the fish body and swim bladder.

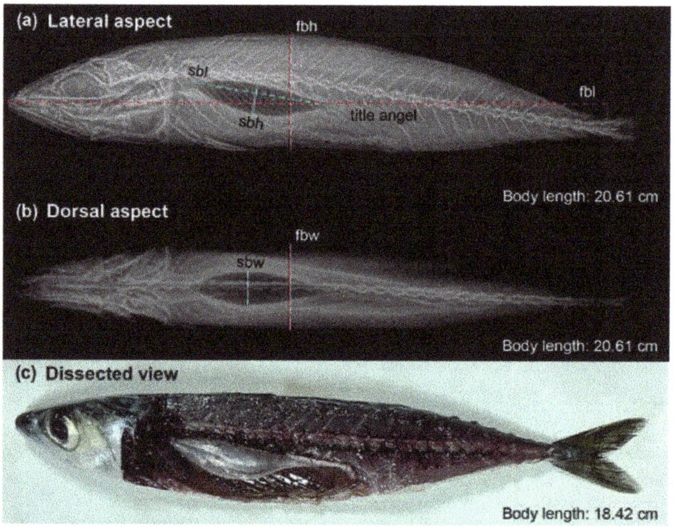

Figure 1. Morphological measurements of Chub mackerel on (**a**) lateral X-ray image, (**b**) dorsal X-ray image, and (**c**) dissected view image. The red lines indicate measurements of the fish body, and the green lines indicate measurements of the swimbladder.

Morphological characteristics of Chub mackerel body and swimbladder were obtained by dissection and X-ray imaging system, respectively. Due to the lack of measurement equipment onboard, swimbladder morphological parameters were obtained through dissection. The abdomen, lateral muscles, and stomach contents were removed, then fish body and swimbladder morphology were measured using a vernier caliper, as shown in Figure 1c. After the survey, morphological images of the fish body and swimbladder were acquired using an X-ray imaging system in the laboratory. The seawater frozen samples were thawed naturally in cold seawater for about 12 h to minimize any damage on swimbladder shape [32,33]. The defrosted samples with better preserved appearance were selected for measuring body length and weight. Subsequently, each sample was scanned with X-rays from both lateral and dorsal aspects to obtain contour images of the fish body and swimbladder (Figure 1a,b). Image files from X-ray scans were processed using Photoshop (V2021, Adobe Systems Inc., San Jose, CA, USA). The measuring scale was constructed based on the fish body length measurements. The fish was appropriately rotated to ensure a level posture, and the grayscale threshold was adjusted to improve the contrast between the swimbladder and the fish body, facilitating accurate identification of the swimbladder boundary. Then manual measurements of the morphological parameters of both the fish body and swimbladder were performed.

According to the measurement results, we conducted approximate calculations of the volume and the cross-sectional area from the dorsal direction of the fish body and swimbladder [41]:

$$V = 4\pi/3 \cdot (l_a/2) \cdot (l_b/2) \cdot (l_c/2) \quad (1)$$

$$S = \pi \cdot (l_a/2) \cdot (l_b/2) \quad (2)$$

where l_a, l_b, and l_c, respectively, represent the length, width, and height of the fish body or swimbladder (cm).

The equivalent spherical radius [33] was estimated following the equation:

$$r = (sbl \cdot sbw \cdot sbh)^{1/3} \quad (3)$$

where sbl, sbw, and sbh, respectively, represent the length, width, and height of the swimbladder (cm).

As a vital organ controlling the buoyancy, the swimbladder typically follows a simple allometric growth pattern. It can be represented using the linear relationship between the equivalent spherical radius and body length:

$$log_{10}L = a + q \cdot log_{10}r \quad (4)$$

where L is the body length (cm), a is a constant, and q is the allometric exponent. When $0 < q < 1$, it is negative allometric growth, when $q = 1$, it is isometric growth, and when $q > 1$, it is positive allometric growth [42].

2.3. Acoustic Backscattering Model

Based on the morphological parameters measured by the X-ray imaging system of the fish body and swimbladder, the Kirchhoff-Ray mode model (KRM) [43] was constructed to simulate the acoustic scattering characteristics of different fish samples. The KRM model approximates the fish body and swimbladder as a series of contiguous cylinders filled with liquid and gas. It calculates the scattering of each part and then coherently sums them to generate the TS result of individual fish [2]. The typical acoustic parameters of fish and water required by the model used were sourced from the research of Clay and Horne [44], listed in Table 1. Using the KRM model, the backscattering cross-sections (σ_{bs}) of fish at different frequencies and incident angles of sound waves were obtained, which is a linear representation of the TS. The frequency range was 1–300 kHz, encompassing the common frequencies used for resource surveys. Since the KRM model becomes inaccurate at a high off-broadside angle [21], the incident wave was limited to the range of 40° to 140°, where the angle greater than 90° is head down, and less than 90° is head up. Referring to the research of Tong et al. [45], the KRM model was established in MATLAB (R2018a, MathWorks Inc., Natick, MA, USA).

Table 1. Acoustic parameters used in the KRM model.

Model Parameters	Values	Unit
Density of sea water	1030	kg/m^3
Density of fish body	1070	kg/m^3
Density of swimbladder	1.24	kg/m^3
Sound speed in sea water	1490	m/s
Sound speed in fish body	1570	m/s
Sound speed in swimbladder	345	m/s

2.4. Ex Situ Target Strength Measurements

The ex situ TS measurements were conducted in an anechoic tank (length, 15 m; width, 7 m; depth, 6 m) at the Fishery Machinery and Instrument Research Institute, Chinese Academy of Fishery Sciences, in October 2023. The measurement system is shown in Figure 2. Two transducers were installed on one side of the tank near the wall, with Mechanism A used to control their position and switch between the transducers. Fish were tethered on the other side of the tank using fishing lines and connected to Mechanism B. The metal rod was suspended below the fish to maintain its stable posture. The distance between the fish and transducers was 7 m, and they were positioned at the same depth of 2 m underwater, with the dorsal side of the fish facing the working transducer. Mechanism B controlled the horizontal rotation of the fish, with angles ranging from 40° (head up) to 140° (head down) at 5° intervals. After each angle rotation, the fish was centered in the beam and held still for approximately 5 s to stabilize its posture, following which echo data were recorded for 5 s. Corresponding to different transducers, the fish underwent the same angle rotations individually. Before and after the experiment, measurements of water temperature and salinity were conducted for transducer calibration. Prior to the start of measurements, the fish were slowly thawed for about 12 h. Then X-ray images were taken, and fish samples undamaged in swimbladder morphology were selected, along with measurements of fish body length and weight.

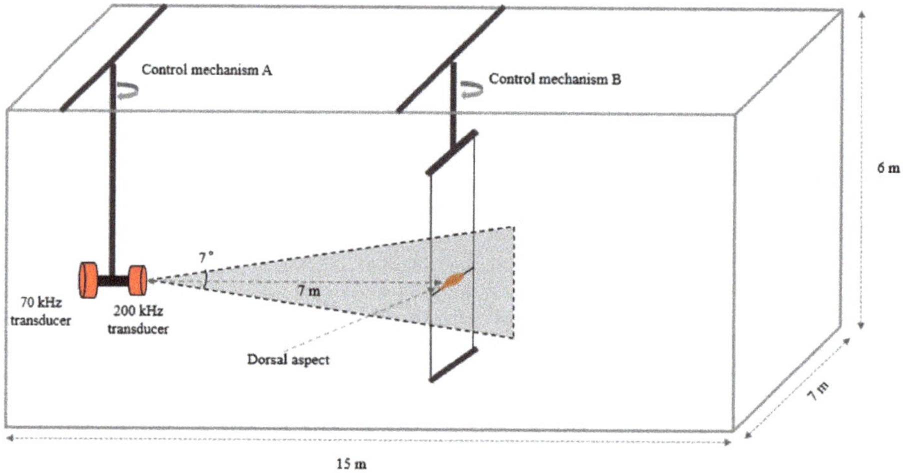

Figure 2. The ex situ TS measurement system. Control mechanism A connects two broadband transducers (center frequency: 70 kHz and 200 kHz), with the gray area representing the beam coverage range. Control mechanism B suspends the fish for measurement, positioning its dorsal side toward the active transducer, while maintaining the fish at the same depth as the transducers.

The Simrad EK80 scientific echosounder system (Kongsberg Maritime AS, Kongsberg, Norway), equipped with two broadband transducers covering frequency from 45 to 90 kHz and 160 to 260 kHz, was used for ex situ TS measurements (Table 2). The two transducers were calibrated using a 38.1 mm diameter tungsten carbide sphere before measurements, following standard methods described by Demer et al. [46]. With reference to the TS spectral response curve of the calibration sphere [47], the frequency bands containing peaks and nulls in the calibration results were excluded from the analysis. Since accurate TS measurements should be taken at least twice the distance beyond the near-field range [48], the near-field ranges corresponding to the two transducers were calculated separately as follows [49]:

$$R = \frac{\pi D^2}{\lambda} \quad (5)$$

where λ is the wavelength (cm) and D is the diameter of the transducer (cm). The transducer calibration and echo data collection were performed using the EK80 software (release V21.15.2, Kongsberg Maritime AS, Kongsberg, Norway) that accompanies the EK80 system.

Table 2. Settings for the EK80 scientific echosounder system.

Parameters	ES70-C	ES200-C
Center frequency (kHz)	70	200
Bandwidth (kHz)	45–90	160–260
Transmitted mode	LFM	LFM
Ramping mode	Fast	Fast
Pulse duration (ms)	1.024	1.024
Sampling frequency (kHz)	62.5	187.5
Beam width (°)	7	7
Transmitted power (W)	750	150
Near-field (m)	2.34	1.07

2.5. Data Analysis

Due to the small sample size and non-normal distribution of the data, we employed the Mann–Whitney U test to compare differences between fish samples measured onboard (before freezing) and those measured by X-ray scans after the survey (after freezing). The Mann–Whitney U test is a non-parametric statistical method used to assess significant differences between two independent groups [50,51]. One advantage of this method is that the two samples do not need to have the same amount of data, and there are no distributional assumptions required that do not assume equal variances between the groups being compared. It is widely applied in diverse fields including medicine, fisheries, and aquaculture [52,53].

To obtain the average TS values of fish targets, the method proposed by Foote [8] was referenced. The average backscattering cross-section (σ_{bs_ave}), which is a linear representation of the average TS, was computed using the Probability Density Function (PDF) of the fish swimming angle distribution by the following:

$$\sigma_{bs\_ave} = \int_{\theta_2}^{\theta_1} \sigma_{bs\_\theta} f(\theta) d\theta \quad (6)$$

$$\mathrm{TS_{ave}} = 10\log_{10} \sigma_{\mathrm{bs\_ave}} \quad (7)$$

Here, θ represents the fish swimming angle, ranging from 40–140°. σ_{bs_θ} represents the backscattering cross-section at angle θ. The fish swimming angle is assumed to follow a normal distribution and $f(\theta)$ represents the PDF of this distribution. In our study, the swimming angle distribution of Chub mackerel was set to N(93°, 4°), as sourced from Nauen et al. [54].

For the KRM model results, the scattering cross-section data can be directly obtained. However, for ex situ TS measurement results, we used the Echoview software (V13.0, Echoview Software Pty Ltd., Tasmania, Australia) to process echo data from the EK80 system. The extraction of TS data at each angle was implemented based on the single target detection-wideband operator provided by the Echoview. The parameter settings are shown in Table 3. The extracted results were then linearly transformed to obtain acoustic scattering cross-section data.

Table 3. Parameters for the single target detection-wideband operator in Echoview.

Parameters	Values	Unit
Operator	Single target detection-wideband	
TS threshold	−80	dB
Pulse length determination level	3/6/9	dB
Minimum normalized pulse length	0.5	
Maximum normalized pulse length	1.5	
Beam compensation model	Simrad Lobe	
Maximum beam compensation	12	dB

In order to investigate the TS characteristics of Chub mackerel, the ratio of the object size to the wavelength was used (L/λ), so that only the effect of changes in acoustic frequency on the TS was considered [55].

Considering that current acoustic estimation of resource abundance still relies mainly on narrowband typical frequencies, the TS–L (body length) relationships were fitted separately using the least-squares method at frequencies of 38 kHz, 70 kHz, 120 kHz, and 200 kHz, referencing empirical formulas [48]:

$$TS = m\log_{10} L + n \tag{8}$$

where m is the slope of the regression, n is the intercept, L is the body length of the fish (cm). The TS of fish with swimbladder is approximately proportional to the square of body length [56], typically with a fixed value of 20 for m, while n is standardized by the square of body length TS_{cm} (b_{20}), as represented by the following equation:

$$TS = 20\log_{10} L + TS_{cm} \tag{9}$$

3. Results

3.1. Swimbladder Morphology and Target Strength Changes before and after Freezing

A total of 37 fresh Chub mackerel samples (before freezing) were dissected and measured for body and swimbladder morphology in the laboratory onboard the research vessel. For the frozen samples, after thorough thawing (about 12 h), X-ray images of 22 Chub mackerel samples were taken and the morphological measurements of the fish body and swimbladder were conducted using Photoshop software. The samples were not the same before and after freezing due to the damage caused by dissection onboard. Both sets of samples were randomly selected.

Before freezing, the length range of samples was 13.3–30.5 cm, with an average length of 18.74 $\pm$ 4.22 cm, and the swimbladder length ranged from 3.01 to 7.27 cm, with an average length of 4.62 $\pm$ 1.19 cm. For the thawed samples measured, the length range was 14.75–26.93 cm, with average length of 20.51 $\pm$ 3.61 cm, the swimbladder length ranged from 2.09 to 7.61 cm, with an average length of 4.79 $\pm$ 1.55 cm, and the angle between the swimbladder and the fish body ranged from 6.1 to 13.1°, with an average tilt angle of 10.10 $\pm$ 1.19° (Table 4).

To investigate the variations in swimbladder morphology before and after freezing, the percentage of swimbladder relative to corresponding parts of the fish body was calculated. The Mann–Whitney U test was then applied to examine significant differences ($p < 0.05$). The results showed that the proportion of the swimbladder in the fish body decreased after freezing. Specifically, there were significant differences ($p < 0.01$) in the ratio of swimbladder height to body height and swimbladder volume to body volume before and after freezing. This indicated that after freezing, there were significant changes in the swimbladder height and volume relative to the fish body. However, the differences in swimbladder length, width, and dorsal cross-sectional area were not significant.

Table 4. Morphological parameters before and after freezing of Chub mackerel and Mann–Whitney U test.

		Before Freezing (Mean ± sd)	After Freezing (Mean ± sd)	Mann–Whitney U test	
				U	p
	No.	37	22		
fish swimbladder	sbl	4.62 ± 1.19	4.79 ± 1.55		
	sbw	0.99 ± 0.35	1.21 ± 0.30		
	sbh	1.01 ± 0.33	1.00 ± 0.26		
	r	1.65 ± 0.47	0.98 ± 0.23		
	Vsb	2.95 ± 2.67	3.38 ± 2.31		
	Ssb	3.82 ± 2.18	4.73 ± 2.39		
fish body	fbl	18.74 ± 4.22	20.51 ± 3.61		
	fbw	2.21 ± 0.64	2.92 ± 0.69		
	fbh	3.10 ± 0.94	4.02 ± 0.90		
	Vfb	81.83 ± 78.05	141.96 ± 90.15		
	Sfb	34.46 ± 18.99	48.88 ± 19.94		
fish swimbladder to fish body ratio	sbl/fbl	0.25 ± 0.03	0.23 ± 0.05	449	
	sbw/fbw	0.45 ± 0.12	0.42 ± 0.09	436	
	sbh/fbh	0.34 ± 0.10	0.25 ± 0.06	622	***
	Vsb/Vfb	0.04 ± 0.02	0.03 ± 0.01	578	**
	Ssb/Sfb	0.11 ± 0.04	0.10 ± 0.03	508	

Note: **: $p < 0.01$, significant differences; ***: $p < 0.001$, extremely significant differences.

The length, volume, and dorsal cross-sectional area of the swimbladder increased significantly with increasing body length ($p < 0.001$). Before and after freezing, the trends of swimbladder volume and dorsal cross-sectional area increasing with body length are generally consistent, while the trend of swimbladder length increasing with body length is slightly higher after freezing compared to before (Figure 3).

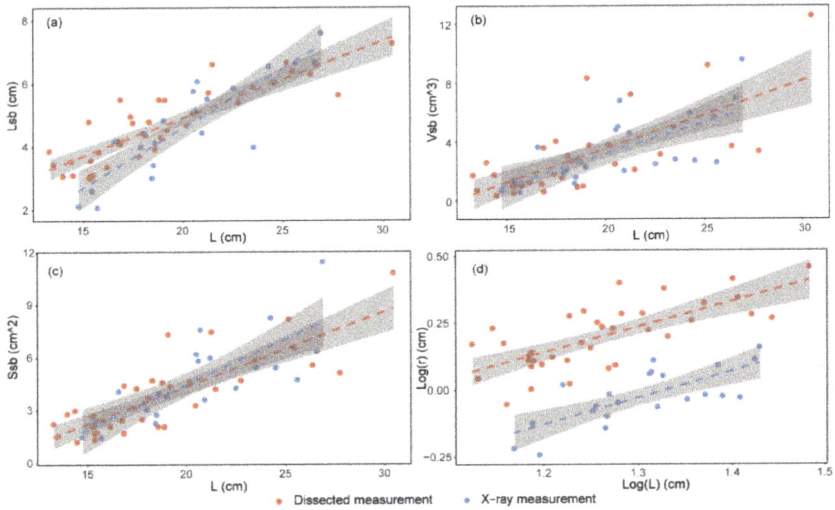

Figure 3. Relationship between the fish body length and the (**a**) swimbladder length, (**b**) swimbladder volume, (**c**) swimbladder cross–sectional area, (**d**) swimbladder equivalent spherical radius. The scatter points represent individual measurements, with red dots indicating fresh sample dissections before freezing and blue dots representing measurements by X-ray after freezing. For all relationships shown in the figures: p–value < 0.001.

The logarithmic form of the swimbladder equivalent radius (r) relative to body length can be used to observe the growth trend of the swimbladder. By establishing the following fitting relationships, it was concluded that the allometric exponent of the swimbladder of Chub mackerel before freezing was 0.945, and after freezing was 0.997, indicating that the swimbladder and the body were in a nearly isometric growth state. It is worth noting that the freezing process itself may cause alterations or damage to the swimbladder structure.

Before freezing:

$$\log_{10} r = 0.945\log_{10} L - 0.992 \tag{10}$$

After freezing:

$$\log_{10} r = 0.997\log_{10} L - 1.32 \tag{11}$$

Based on post–freezing X-ray imaging data and the differences in swimbladder proportions within the fish body before and after freezing, we used the KRM model to estimate the TS data of 22 fish samples both before and after freezing. The KRM model angle range is 40–140°, and the parameters required for the model, such as those for the fish body, swimbladder, and water, are shown in Table 1. Additionally, the model assumes consistent environmental conditions and homogeneous material properties. As shown in Figure 4, ΔTS represents the difference between the TS before and after freezing. The results showed that despite the reduction in the swimbladder proportion within the fish body due to freezing, the TS did not exhibit a clear trend of increase or decrease at the four typical frequencies of 38, 70, 120, and 200 kHz. The average TS differences at these frequencies were: 0.08 ± 0.19 dB, 0.06 ± 0.30 dB, 0.02 ± 0.78 dB, and 0.23 ± 0.60 dB, respectively. As the frequency increased, the difference in TS before and after freezing gradually became more pronounced, indicating a higher sensitivity of TS changes at higher frequencies.

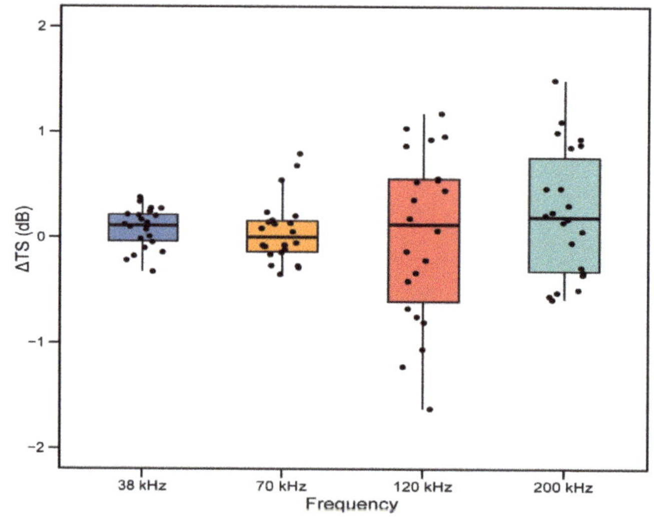

Figure 4. Differences in TS before and after freezing at typical frequencies (38 kHz, 70 kHz, 120 kHz, and 200 kHz) based on the KRM model. ΔTS represents the difference between the TS before and after freezing.

3.2. Comparison of Target Strength between Ex Situ Experiment and KRM Model

Seven individuals were randomly selected from the collected samples (TL: 20.74 ± 1.11 cm) and after thorough thawing for 12 h, KRM model simulation and ex situ TS measurements were conducted to compare the differences between the two methods. X-ray imaging was performed before and after the ex situ measurements to examine the integrity of the swimbladder morphology.

Figure 5 shows the variation in the TS spectra of single Chub mackerel (TL: 21.2 cm) obtained from KRM model simulation and ex situ measurements. Overall, changes in TS relative to angle and frequency were very clearly observed for both measurements. At different frequencies, the maximum TS values were distributed around 100°, when the fish head was away from the transducer and the swimming posture was downward. Combining the measured data of the swimbladder tile angle (10.10° ± 1.49°), the swimbladder was perpendicular to the incident angle of the sound wave, indicating the maximum cross-sectional relative to the sound wave. As frequency increased, the TS exhibited more pronounced directionality. Compared with the regular changes in the TS spectra of the KRM model, the ex situ measurement results showed more heterogeneous TS variations in certain angular intervals.

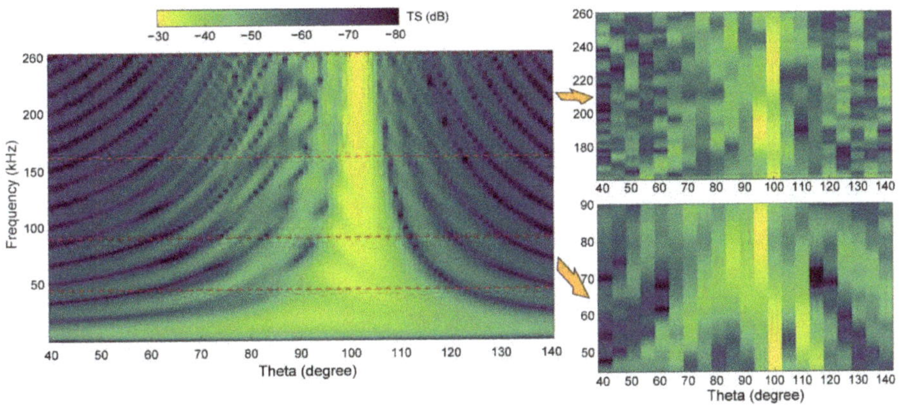

Figure 5. TS variation with angle and frequency for an individual Chub mackerel (body length: 21.2 cm). On the left: the KRM model estimation results, frequency range from 1 to 260 kHz; on the right: ex situ measurement results using broadband transducers for 45–90 kHz and 160–260 kHz frequency ranges. The horizontal axis represents the angle of the fish body relative to the detection beam, ranging from 40 to 140°.

Figure 6 shows the average TS differences of Chub mackerel at various frequencies between the KRM model and ex situ measurements (ΔTS: $TS_{ex\text{-}situ}-TS_{KRM}$), selecting frequency bands within the effective calibration range (45–80 kHz, 180–200 kHz, and 230–240 kHz). The results indicated that the differences between KRM model and ex situ measurements were concentrated within ±3 dB, with an average difference of −0.38 ± 1.84 dB. At the edges of the frequency bands, such as 47 kHz, 80 kHz, 230 kHz, and 233 kHz, larger differences were observed, with some frequencies showing differences exceeding 3 dB, although the larger differences were only found in a single individual case.

3.3. Broadband Scattering Response Characteristics

A total of 22 X-ray scan datasets of Chub mackerel with well-preserved swimbladders were obtained, and the broadband scattering response characteristics were investigated using the KRM model. The average TS of individual fish increased rapidly in the range of approximately 1–30 kHz, which corresponds to the resonance frequency range. Subsequently, as the frequency increased, the change in average TS gradually slowed down and tended to stabilize. There were certain differences in individual TS due to variations in the length of fish samples. By calculating the average TS of all individuals, it was observed that when the frequency exceeded 100 kHz, the average TS remained relatively stable, around −40.76 dB. The maximum TS gradually increased with frequency, and the trend began to stabilize when the frequency reached 150 kHz. As the frequency increased, the difference between the maximum TS and average TS also increased, corresponding to the directional characteristics of TS. At lower frequencies, the differences in TS between angles were not significant, but as the frequency increased, the directional characteristics of TS became more apparent, leading to larger differences in TS between angles (Figure 7).

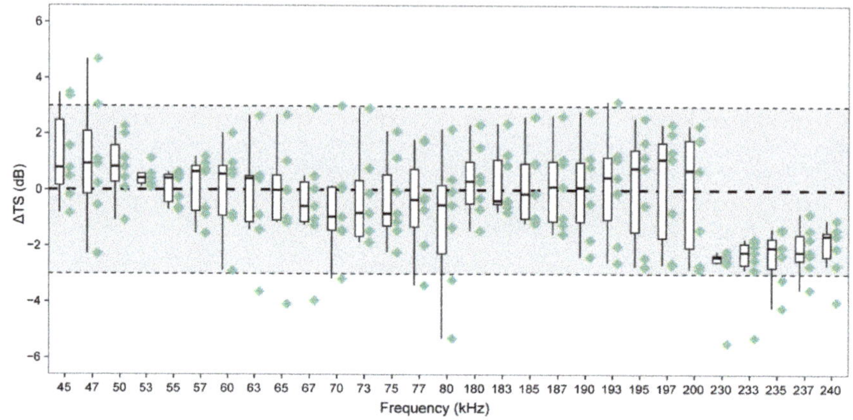

Figure 6. The difference in average TS between Chub mackerel ex situ measurements and KRM model estimates. The horizontal axis represents different frequencies, ranging from 45 to 80 kHz, 180 to 200 kHz, and 230 to 240 kHz. The vertical axis indicates the difference between $TS_{ex\ situ}$ and TS_{KRM}. The shaded area represents the range of ±3 dB.

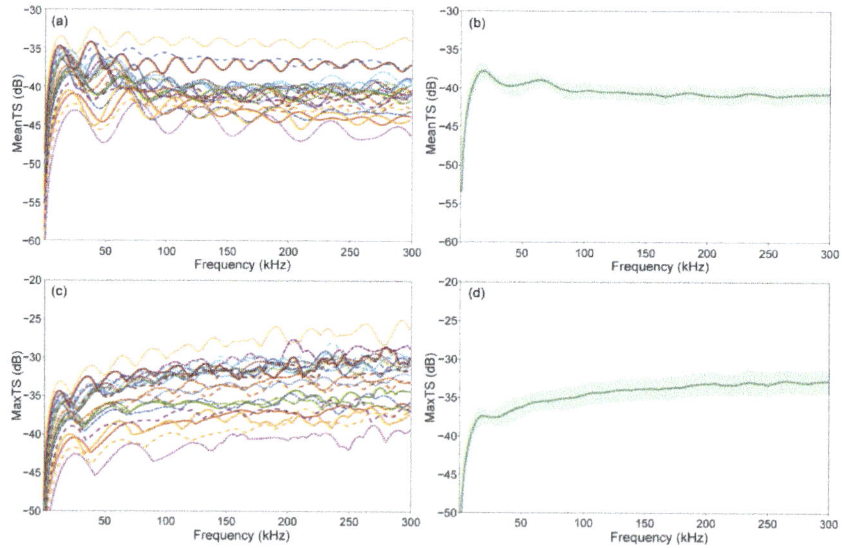

Figure 7. Based on the KRM model results, the average and maximum TS of Chub mackerel vary with frequency: (**a**,**c**) represent the average and maximum TS for different individual fish, with different colored lines representing different individuals; (**b**,**d**) show the average values of average and maximum TS.

We used the ratio of body length (L) to wavelength (λ, c/f) to simultaneously examine the variation of Chub mackerel *TS* with body length and frequency (Figure 8). Within the scope of the study, individual average TScm values were relatively concentrated, fluctuating within a range of about 5 dB, and the fluctuation became smoother as L/λ increased. Then the TScm of all individuals was averaged, which could reflect the trend of TScm more clearly, and when $L/\lambda > 10$, the TScm basically did not fluctuate. Therefore, the TScm was averaged over all individual TScm in this range ($30 > L/\lambda > 10$) to obtain the relationship between Chub mackerel *TS* and body length under the broadband conditions:

$$TS = 20\log_{10} L - 66.76 \tag{12}$$

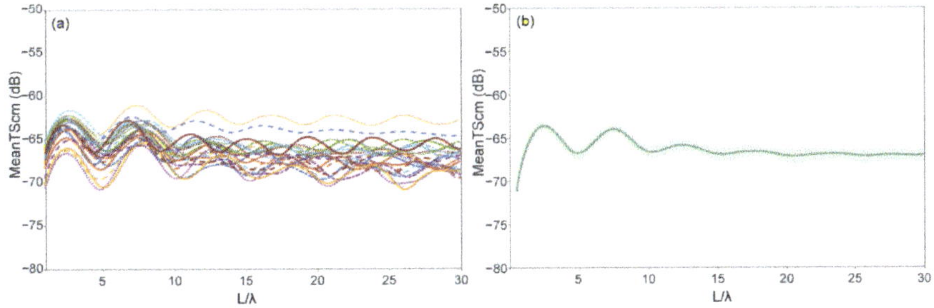

Figure 8. Based on the KRM model results, the relationship between average TScm and L/λ for Chub mackerel: (**a**) represents the average TScm for different individuals of Chub mackerel, with different colored lines representing the variations in TScm among different individuals; (**b**) represents the mean value of average TScm for all individuals of Chub mackerel, with the shaded areas indicating the 95% confidence intervals.

3.4. Relationships between Target Strength and Body Length at Typical Frequencies

Currently, marine fishery resource surveys mainly rely on typical narrowband frequencies such as 38 kHz, 70 kHz, 120 kHz, and 200 kHz. Based on the results of the KRM model, we fitted the TS–L relationships at different typical frequencies to support the acoustic estimation of fishery resources (Figure 9). The specific empirical formulas are as follows.

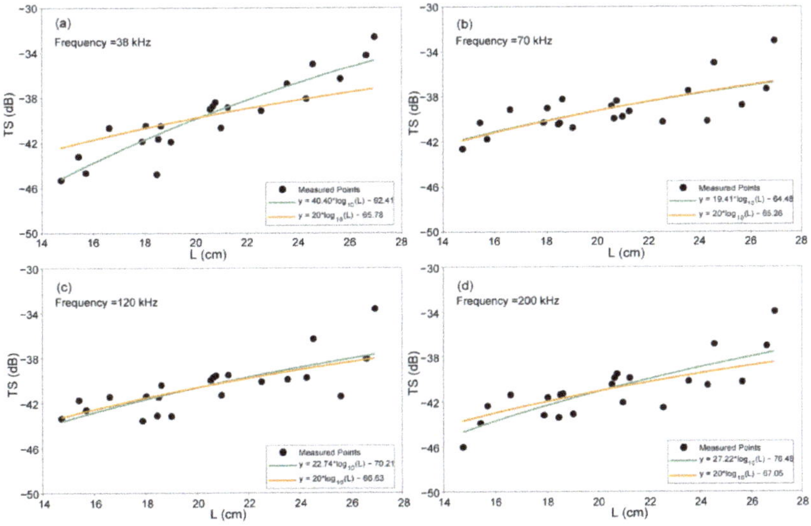

Figure 9. The relationship between TS and body length (L) for Chub mackerel at 38 kHz, 70 kHz, 120 kHz, and 200 kHz derived from the KRM model.

38 kHz:

$$TS = 40.4\log_{10} L - 92.41 \tag{13}$$

(95% CI: m (32.14, 48.65), n (−103.20, −81.62), R^2: 0.84)

$$TS = 20\log_{10} L - 65.78 \tag{14}$$

(95% CI: TScm (−66.70, −64.86), R^2: 0.63)

70 kHz:

$$TS = 19.41\log_{10} L - 64.48 \tag{15}$$

(95% CI: m (10.34, 28.47), n (−76.33, −52.63), R^2: 0.50)

$$TS = 20\log_{10} L - 65.26 \tag{16}$$

(95% CI: TScm (−65.92, −64.59), R^2: 0.50)

120 kHz:

$$TS = 22.74\log_{10} L - 70.21 \tag{17}$$

(95% CI: m (13.18, 32.30), n (−82.72, −57.71), R^2: 0.55)

$$TS = 20\log_{10} L - 66.63 \tag{18}$$

(95% CI: TScm (−67.34, −65.93), R^2: 0.54)

200 kHz:

$$TS = 27.22\log_{10} L - 76.48 \tag{19}$$

(95% CI: m (17.68, 36.76), n (−88.96, −64.01), R^2: 0.64)

$$TS = 20\log_{10} L - 67.05 \tag{20}$$

(95% CI: TScm (−67.79, −66.32), R^2: 0.59)

4. Discussion

In the present study, we measured changes in swimbladder morphology before and after freezing preservation and examined the effects on TS. We compared two common methods for TS research in pelagic fish: ex situ measurements and model estimations, validating the applicability of the KRM model. Based on the KRM, the broadband scattering characteristics of Chub mackerel and the relationship between TS and body length were derived. The results of this study could facilitate species identification and resource assessment of Chub mackerel using acoustic technology, thereby promoting the application of fisheries acoustic technology in the Northwest Pacific Ocean.

For swimbladder-bearing fish, the air stored in the swimbladder creates different media conditions for acoustic pulse transmission, accounting for over 95% of the fish TS [8]. Therefore, studies of TS in swimbladder-bearing fish must consider the effects of the swimbladder. Previous research [12,45] have shown that the variation in fish TS at different angles was consistent with the TS of the swimbladder, with both being very close to each other. The maximum fish TS occurred when the maximum cross-section of the swimbladder was perpendicular to the angle of acoustic incidence. Our results were also consistent with these findings. Additionally, we observed TS characteristics at different angles across successive frequencies and found that TS became more sensitive to angle changes as the frequency increased (Figure 5).

The diel vertical migration (DVM) of small pelagic fish such as Chub mackerel, which accompanies the plankton in the scattering layer, may result in changes in the volume of their swimbladder [57]. When fish ascend, the water pressure decreases, causing the fish to relax and the swimbladder to expand. Conversely, when fish descend, the water pressure increases, causing the fish to tense up and the swimbladder to contract. We considered the changes in swimbladder morphology associated with this fish behavior, and biological sampling was primarily conducted at night in shallow water layers to minimize changes in swimbladder morphology due to water pressure. Tong et al. [45] used Boyle's law to model the variation of Chub mackerel TS at different depths, but there is a lack of actual data to support the simulation. Future research could consider obtaining live Chub mackerel samples and simulating different water pressure in the laboratory to study changes in the swimbladder volume in different water layers and the effects on TS.

The swimbladder structure of the most economically important pelagic fish is similar to that of the Chub mackerel, with the swimbladder changing according to the depth at which they live, as they are all physostomous fish. In contrast, some mesopelagic and bathypelagic fish, often linked to a deep scattering layer, lack a connection between the swimbladder and the alimentary canal, causing the swimbladder volume to remain constant regardless of water depth [58,59]. Small numbers of epipelagic and demersal fish, including Skipjack tuna (*Katsuwonus pelamis*) and Atlantic mackerel (*Scomber scombrus*), lack a swimbladder. And their acoustic scattering mainly comes from their flesh and backbone, making it much weaker compared to swimbladder-bearing fish [60]. The presence or absence of a swimbladder can be used as a criterion for distinguishing these different marine organisms based on acoustic methods. Although some of these organisms have no global economic value, they are an indispensable part of the marine ecosystem. Separating these from economically significant fisheries resources facilitates the conservation of living marine resources and the maintenance of ecosystems.

Ideally, fresh samples would be preferred for studies of target strength, but this is more challenging for most pelagic organisms. Many studies have used the same method as ours, with rapid freezing in seawater at temperatures below −20 °C. Yasuma et al. [33] pointed out that this procedure has a negligible effect on the swimbladder. However, other studies [61] have indicated that freezing can indeed cause damage to the shape or structure of the fish muscle, which in turn affects the swimbladder. Freezing can also increase the number of bubbles on the surface of the fish, resulting in an increase in TS by 1.9–9.8 dB [62]. There is still a lack of comprehensive research on the effects of different preservation methods on the swimbladder. This study excluded swimbladder changes caused by fish behavior and quantified the effects of seawater freezing preservation on swimbladder morphology. We selected fish individuals that exhibited greater activity and showed no signs of external deformation for sampling. After placing them in seawater for approximately 1 h, we selected samples for dissection or to be frozen, thus minimizing any potential interference with swimbladder morphology during the sampling process. Due to the absence of non-destructive instruments such as X-ray machines on the research vessel, we were unable to compare the same samples before and after freezing. Instead, we minimized potential research errors caused by sample variability by assessing the proportion of swimbladder to different parts of the fish body and employing random sampling. The results showed significant differences in swimbladder height and volume before and after freezing, but those changes did not lead to a consistent increase or decrease in TS. Therefore, we recommend that future research should preferably utilize fresh samples for constructing acoustic models or conducting TS measurements, especially under high-frequency conditions.

By utilizing morphological parameters of fish body and swimbladder, constructing acoustic scattering models for individual fish enables simulation and calculation of TS. This method offers the advantages of flexibility, simplicity in computation, and independence from external environmental constraints. Moreover, with continuous technological advancements, the accuracy of model assessments has improved. Fish TS research based on modeling methods has been applied in various studies on the scattering characteristics of fish both domestically and internationally [26,62,63]. The KRM model, as a physical approximation model, approximates the fish body and swimbladder as multiple contiguous cylinders. It calculates the TS of each individual cylindrical component and then sums them up to obtain the overall TS value of the entire fish [21]. However, fish TS characteristics derived solely from models often require comparison with measurements conducted ex situ or in situ for validation [12,64]. We compared the KRM model of Chub mackerel with ex situ TS measurement results. When used to study changes in TS at different angles and frequencies, the KRM model not only exhibited significant directional characteristics but also showed more regular TS variations. In contrast, ex situ measurements showed more erratic changes at certain angles. By calculating the average TS at different frequencies, we further demonstrated the applicability of the KRM model in modeling Chub mackerel TS studies. The KRM model assumes that $ka > 0.15$ (where a is the minor radius of a cylinder and k is the wavenumber) for each cylinder derived from the fish body and swimbladder, making the model less accurate when the sound waves are close to the main axis of the fish body [55]. The accuracy of the KRM model is also influenced by the digitization of fish body morphology, including the precision of segmenting the swimbladder and fish body. Additionally, the speed of sound and density values of the fish body, swimbladder, and water are crucial parameters in the KRM model [45]. In our study, we used relevant parameters from a general KRM model for marine fish. It is essential to perform actual measurements of these key parameters in future related experiments to improve the model's accuracy.

In acoustic surveys of fishery resources, obtaining accurate TS–L relationships is crucial for accurately estimating fish stock abundance. In this study, we established empirical relationships between TS and L for Chub mackerel samples with a length range of 14.75–26.93 cm under commonly used fishery survey frequencies. We also compared the previous studies reported in the literature. At the four typical frequencies, the values of TScm were closer to those of Park et al. [23] based on the KRM model (16–28 cm), with differences of 0.24, 1.24, −0.63, and 0.30 dB at each frequency. However, using the same model, the TScm values reported by Tong et al. [45] (12–22 cm) were generally 6–8 dB lower than our results, which may be attributed to differences in body length distribution. Zhu et al. [65] conducted ex situ TS measurements of free-swimming Chub mackerel (17.4–34 cm) in a net cage, and the difference in TScm was 2.12 dB at 38 kHz and 2.57 dB at 120 kHz, which was mainly related to the different angular distributions of the swimming postures, and the TS changes caused by the posture angle were more pronounced with the increase in frequency.

The commercial use of scientific echosounder systems such as the Simrad EK80 system, which simultaneously acquire narrowband and broadband echo data, has shifted research focus toward broadband types [66]. The application of broadband technology in acoustic surveys of fishery resources has become a research hotspot, and studies based on acoustic scattering models can provide reference for species identification based on broadband characteristics. In this study, we utilized the KRM scattering model to construct TS spectral features. Compared to other research, Yan et al. [67] measured the broadband TS of three swimbladder-less fish species, and Lucca et al. [68] measured the broadband TS of various shrimp and shelled pteropods species. These different species exhibited distinct spectral characteristics, and these differences are key to species identification based on broadband acoustic technology.

Our study has several limitations: Firstly, the samples measured before and after freezing were not the same individuals, and the sample sizes differed, which may lead to uncertainties in TS changes. Secondly, different methods were used for measuring the swimbladder, which introduces uncertainties in TS changes based on these differences. The parameters used in the KRM model were mainly derived from previous literature, without on-site measurements. Additionally, the sample size was limited, and the geographic scope was narrow. To address these issues, we plan to use the same batch and a consistent number of samples for pre- and post-freezing comparisons in future research, adopt consistent methods for measuring the swimbladder, and conduct on-site parameter measurements. Furthermore, we will increase the geographic range of our samples to improve the applicability of our results. We believe that addressing these limitations will enhance the contribution of our study to marine acoustic research.

5. Conclusions

This study investigated the impact of freezing preservation on the swimbladder morphology and the TS of Chub mackerel, and compared two methods for TS estimation, then examined the broadband scattering characteristics. Our findings indicated that freezing preservation leads to significant changes in swimbladder morphology, notably in its height and volume. However, these morphological changes do not result in consistent and significant variations in TS. The comparison between the KRM model and ex situ TS measurements revealed a good agreement, with differences concentrated within ± 3 dB and an average difference of -0.38 ± 1.84 dB. This validated the applicability of the KRM model for TS estimation of Chub mackerel under various conditions. Based on the KRM results, the TS of Chub mackerel exhibited significant directivity, with fluctuations gradually decreasing and stabilizing as frequency increased. In the broadband mode, the relationship between TS and body length (L) of Chub mackerel was $TS = 20\log(L) - 66.76$ $(30 > L/\lambda > 10)$. Our research could provide essential data for acoustic resource estimation and species identification.

Author Contributions: Conceptualization, Z.Z.; Funding acquisition, B.L.; Methodology, Z.Z., J.T., and M.X.; Project administration, J.T. and B.L.; Software, Z.Z., M.X., C.Q., and S.L.; Supervision, J.T.; Writing—original draft, Z.Z.; Writing—review and editing, J.T. All authors have read and agreed to the published version of the manuscript.

Funding: This research was funded by the National Key R&D Program of China (2023YFD2401302, 2019YFD0901401). We also acknowledge funds provided by the Ministry of Agriculture and Rural Affairs of China through the project on the Survey and Monitor-Evaluation of Global Fishery Resources.

Institutional Review Board Statement: The study was approved by the Shanghai Ocean University Scientific Research Department (Approval code: SHOU-DW-2023-097).

Informed Consent Statement: Not applicable.

Data Availability Statement: Data will be made available on request.

Acknowledgments: The authors of this research would like to thank all the researchers and sailors on RV Songhang, especially Shujie Wan and Hao Xv, who contributed to the data collection during the marine survey. Special thanks to the Fishery Machinery and Instrument Research Institute, Chinese Academy of Fishery Sciences, for providing experimental site support. Thanks to the Institute of Hydroecology, MWR, and CAS for providing equipment support. Thanks to the reviewers for the thoughtful discussion that substantially improved the original manuscript.

Conflicts of Interest: The authors declare no conflicts of interest.

References

1. Andersen, L.N.; Chu, D.; Handegard, N.O.; Heimvoll, H.; Korneliussen, R.; Macaulay, G.J.; Ona, E.; Patel, R.; Pedersen, G. Quantitative Processing of Broadband Data as Implemented in a Scientific Split-Beam Echosounder. *Methods Ecol. Evol.* **2024**, *15*, 317–328. [CrossRef]
2. Palermino, A.; De Felice, A.; Canduci, G.; Biagiotti, I.; Costantini, I.; Centurelli, M.; Leonori, I. Application of an Analytical Approach to Characterize the Target Strength of Ancillary Pelagic Fish Species. *Sci. Rep.* **2023**, *13*, 15182. [CrossRef] [PubMed]
3. Sun, M.; Cai, Y.; Zhang, K.; Zhao, X.; Chen, Z. A Method to Analyze the Sensitivity Ranking of Various Abiotic Factors to Acoustic Densities of Fishery Resources in the Surface Mixed Layer and Bottom Cold Water Layer of the Coastal Area of Low Latitude: A Case Study in the Northern South China Sea. *Sci. Rep.* **2020**, *10*, 11128. [CrossRef] [PubMed]
4. Simmonds, J.; MacLennan, D.N. *Fisheries Acoustics: Theory and Practice*; John Wiley & Sons: Hoboken, NJ, USA, 2008.
5. Stanton, T.K.; Sellers, C.J.; Jech, J.M. Resonance Classification of Mixed Assemblages of Fish with Swimbladders Using a Modified Commercial Broadband Acoustic Echosounder at 1–6 kHz. *Can. J. Fish. Aquat. Sci.* **2012**, *69*, 854–868. [CrossRef]
6. Andersen, L.N.; Chu, D.; Heimvoll, H.; Korneliussen, R.; Macaulay, G.J.; Ona, E. Quantitative Processing of Broadband Data as Implemented in a Scientific Splitbeam Echosounder. *arXiv* **2021**, arXiv:2104.07248. [CrossRef]
7. Yamamoto, N.; Amakasu, K.; Abe, K.; Matsukura, R.; Imaizumi, T.; Matsuura, T.; Murase, H. Volume Backscattering Spectra Measurements of Antarctic Krill Using a Broadband Echosounder. *Fish. Sci.* **2023**, *89*, 301–315. [CrossRef]
8. Foote, K.G. Importance of the Swimbladder in Acoustic Scattering by Fish: A Comparison of Gadoid and Mackerel Target Strengths. *J. Acoust. Soc. Am.* **1980**, *67*, 2084–2089. [CrossRef]

9. Hazen, E.L.; Horne, J.K. A Method for Evaluating the Effects of Biological Factors on Fish Target Strength. *ICES J. Mar. Sci.* **2003**, *60*, 555–562. [CrossRef]
10. Lu, H.-J.; Kang, M.; Huang, H.-H.; Lai, C.-C.; Wu, L.-J. Ex Situ and in Situ Measurements of Juvenile Yellowfin Tuna Thunnus Albacares Target Strength. *Fish. Sci.* **2011**, *77*, 903–913. [CrossRef]
11. Pérez-Arjona, I.; Godinho, L.; Espinosa, V. Influence of Fish Backbone Model Geometrical Features on the Numerical Target Strength of Swimbladdered Fish. *ICES J. Mar. Sci.* **2020**, *77*, 2870–2881. [CrossRef]
12. Madirolas, A.; Membiela, F.A.; Gonzalez, J.D.; Cabreira, A.G.; dell'Erba, M.; Prario, I.S.; Blanc, S. Acoustic Target Strength (TS) of Argentine Anchovy (*Engraulis Anchoita*): The Nighttime Scattering Layer. *ICES J. Mar. Sci.* **2017**, *74*, 1408–1420. [CrossRef]
13. Li, B.; Liu, J.; Gao, X.; Huang, H.; Wang, F.; Huang, Z. Acoustic Target Strength of Thornfish (*Terapon jarbua*) Based on the Kirchhoff-Ray Mode Model. *Electronics* **2024**, *13*, 1279. [CrossRef]
14. Horne, J.K. Acoustic Approaches to Remote Species Identification: A Review. *Fish. Oceanogr.* **2000**, *9*, 356–371. [CrossRef]
15. Sobradillo, B.; Boyra, G.; Martinez, U.; Carrera, P.; Peña, M.; Irigoien, X. Target Strength and Swimbladder Morphology of Mueller's Pearlside (*Maurolicus muelleri*). *Sci Rep.* **2019**, *9*, 17311. [CrossRef] [PubMed]
16. Boyra, G.; Moreno, G.; Orue, B.; Sobradillo, B.; Sancristobal, I. In Situ Target Strength of Bigeye Tuna (*Thunnus obesus*) Associated with Fish Aggregating Devices. *ICES J. Mar. Sci.* **2019**, *76*, 2446–2458. [CrossRef]
17. Hasegawa, K.; Yan, N.; Mukai, T. In Situ Broadband Acoustic Measurements of Age-0 Walleye Pollock and Pointhead Flounder in Funka Bay, Hokkaido, Japan. *J. Mar. Sci. Technol.* **2021**, *29*, 135–145. [CrossRef]
18. Kang, D.; Cho, S.; Lee, C.; Myoung, J.-G.; Na, J. Ex Situ Target-Strength Measurements of Japanese Anchovy (*Engraulis japonicus*) in the Coastal Northwest Pacific. *ICES J. Mar. Sci.* **2009**, *66*, 1219–1224. [CrossRef]
19. Kim, H.; Cho, S.; Kim, M.; Kim, S.; Kang, D. Acoustic Target Strength According to Different Growth Stages of Japanese Anchovy (*Engraulis japonicus*): A Comparison of Juvenile and Adult Fish. *J. Mar. Sci. Eng.* **2023**, *11*, 1575. [CrossRef]
20. Yoon, E.; Lee, H.; Park, C.; Lee, Y.-D.; Hwang, K.; Kim, D.N. Ex Situ Target Strength of Yellow Croaker (*Larimichthys polyactis*) in a Seawater Tank. *Fish. Res.* **2023**, *260*, 106610. [CrossRef]
21. Macaulay, G.J.; Pena, H.; Fassler, S.M.M.; Pedersen, G.; Ona, E. Accuracy of the Kirchhoff-Approximation and Kirchhoff-Ray-Mode Fish Swimbladder Acoustic Scattering Models. *PLoS ONE* **2013**, *8*, e64055. [CrossRef] [PubMed]
22. Bonomo, A.L.; Isakson, M.J. Modeling the Acoustic Scattering from Axially Symmetric Fluid, Elastic, and Poroelastic Objects Due to Nonsymmetric Forcing Using COMSOL Multiphysics. In Proceedings of the 2016 COMSOL Conference, Boston, MA, USA, 5–7 October 2016.
23. Park, G.; Oh, W.; Oh, S.; Lee, K. Acoustic scattering characteristics of chub mackerel (*Scomber japonicus*) by KRM model. *J. Korean Soc. Fish. Technol.* **2022**, *58*, 32–38. [CrossRef]
24. Sawada, K.; Takahashi, H.; Abe, K.; Ichii, T.; Watanabe, K.; Takao, Y. Target-Strength, Length, and Tilt-Angle Measurements of Pacific Saury (*Cololabis saira*) and Japanese Anchovy (*Engraulis japonicus*) Using an Acoustic-Optical System. *ICES J. Mar. Sci.* **2009**, *66*, 1212–1218. [CrossRef]
25. Pérez-Arjona, I.; Godinho, L.; Espinosa, V. Numerical Simulation of Target Strength Measurements from near to Far Field of Fish Using the Method of Fundamental Solutions. *Acta Acust. United Acust.* **2018**, *104*, 25–38. [CrossRef]
26. Yang, H.; Cheng, J.; Tang, T.; Chen, J.; Li, G. Acoustic Target Strength of Jellyfish, *Nemopilema nomurai*, Measured at Multi-Frequency and Multi-Orientation. *J. Appl. Ichthyol.* **2023**, *2023*, 6650863. [CrossRef]
27. Jech, J.M.; Horne, J.K.; Chu, D.; Demer, D.A.; Francis, D.T.; Gorska, N.; Jones, B.; Lavery, A.C.; Stanton, T.K.; Macaulay, G.J. Comparisons among Ten Models of Acoustic Backscattering Used in Aquatic Ecosystem Research. *J. Acoust. Soc. Am.* **2015**, *138*, 3742–3764. [CrossRef] [PubMed]
28. Hazen, E.L.; Horne, J.K. Comparing the Modelled and Measured Target-Strength Variability of Walleye Pollock, *Theragra chalcogramma*. *ICES J. Mar. Sci.* **2004**, *61*, 363–377. [CrossRef]
29. Peña, H.; Foote, K.G. Modelling the Target Strength of *Trachurus symmetricus* Murphyi Based on High-Resolution Swimbladder Morphometry Using an MRI Scanner. *ICES J. Mar. Sci.* **2008**, *65*, 1751–1761. [CrossRef]
30. Sawada, K.; Uchikawa, K.; Matsuura, T.; Sugisaki, H.; Amakasu, K.; Abe, K. In Situ and Ex Situ Target Strength Measurement of Mesopelagic Lanternfish, Diaphus Theta (Family *Myctophidae*). *J. Mar. Sci. Technol.* **2011**, *19*, 10. [CrossRef]
31. Sobradillo, B.; Boyra, G.; Pérez-Arjona, I.; Martinez, U.; Espinosa, V. Ex Situ and in Situ Target Strength Measurements of European Anchovy in the Bay of Biscay. *ICES J. Mar. Sci.* **2021**, *78*, 782–796. [CrossRef]
32. Yasuma, H.; Sawada, K.; Ohshima, T.; Miyashita, K.; Aoki, I. Target Strength of Mesopelagic Lanternfishes (Family *Myctophidae*) Based on Swimbladder Morphology. *ICES J. Mar. Sci.* **2003**, *60*, 584–591. [CrossRef]
33. Yasuma, H.; Sawada, K.; Takao, Y.; Miyashita, K.; Aoki, I. Swimbladder Condition and Target Strength of Myctophid Fish in the Temperate Zone of the Northwest Pacific. *ICES J. Mar. Sci.* **2010**, *67*, 135–144. [CrossRef]
34. FAO. *The State of World Fisheries and Aquaculture*; Food & Agriculture Organization: Rome, Italy, 2022; Volume 3.
35. Cai, K.; Kindong, R.; Ma, Q.; Tian, S. Stock Assessment of Chub Mackerel (*Scomber japonicus*) in the Northwest Pacific Using a Multi-Model Approach. *Fishes* **2023**, *8*, 80. [CrossRef]
36. NPFC Secretariat. Summary Footprint of Chub Mackerel Fisheries. NPFC-2023-AR-Annual Summary Footprint. 2023. Available online: https://www.npfc.int/summary-footprint-chub-mackerel-fisheries (accessed on 1 May 2024).
37. Shi, Y.; Zhang, X.; He, Y.; Fan, W.; Tang, F. Stock Assessment Using Length-Based Bayesian Evaluation Method for Three Small Pelagic Species in the Northwest Pacific Ocean. *Front. Mar. Sci.* **2022**, *9*, 775180. [CrossRef]

38. Sogawa, S.; Hidaka, K.; Kamimura, Y.; Takahashi, M.; Saito, H.; Okazaki, Y.; Shimizu, Y.; Setou, T. Environmental Characteristics of Spawning and Nursery Grounds of Japanese Sardine and Mackerels in the Kuroshio and Kuroshio Extension Area. *Fish Oceanogr.* **2019**, *28*, 454–467. [CrossRef]
39. Xue, M.; Tong, J.; Tian, S.; Wang, X. Broadband Characteristics of Zooplankton Sound Scattering Layer in the Kuroshio–Oyashio Confluence Region of the Northwest Pacific Ocean in Summer of 2019. *J. Mar. Sci. Een.* **2021**, *9*, 938. [CrossRef]
40. Zhu, Z.; Tong, J.; Xue, M.; Sarr, O.; Gao, T. Assessing the Influence of Abiotic Factors on Small Pelagic Fish Distribution across Diverse Water Layers in the Northwest Pacific Ocean through Acoustic Methods. *Eco. Indic.* **2024**, *158*, 111563. [CrossRef]
41. Sarmiento-Lezcano, A.N.; Pilar Olivar, M.; Jose Caballero, M.; Couret, M.; Hernandez-Leon, S.; Castellon, A.; Pena, M. Swimbladder Properties of *Cyclothone* spp. in the Northeast Atlantic Ocean and the Western Mediterranean Sea. *Front. Mar. Sci.* **2023**, *10*, 1093982. [CrossRef]
42. Saenger, R.A. Bivariate Normal Swimbladder Size Allometry Models and Allometric Exponents for 38 Mesopelagic Swimbladdered Fish Species Commonly Found in the North Sargasso Sea. *Can. J. Fish. Aquat. Sci.* **1989**, *46*, 1986–2002. [CrossRef]
43. Clay, C.S.; Horne, J.K. Acoustic Models and Target Strengths of the Atlantic Cod (*Gadus morhua*). *JASA* **1992**, *92*, 2350–2351. [CrossRef]
44. Clay, C.S.; Horne, J.K. Acoustic Models of Fish: The Atlantic Cod (*Gadus morhua*). *JASC* **1994**, *96*, 1661–1668. [CrossRef]
45. Tong, J.; Xue, M.; Zhu, Z.; Wang, W.; Tian, S. Impacts of Morphological Characteristics on Target Strength of Chub Mackerel (*Scomber japonicus*) in the Northwest Pacific Ocean. *Front. Mar. Sci.* **2022**, *9*, 10. [CrossRef]
46. Demer, D.A.; Berger, L.; Bernasconi, M.; Bethke, E.; Boswell, K.; Chu, D.; Domokos, R.; Dunford, A.; Fassler, S.; Gauthier, S.; et al. Calibration of Acoustic Instruments; ICES: 2015. Available online: https://repository.oceanbestpractices.org/handle/11329/626 (accessed on 30 May 2024).
47. Simrad. *Simrad EK80 Wide Band Scientific Echo Sounder Reference Manual, Release: 21.15; Kongsberg Maritime AS., Kongsberg, Norway.* 2022. Available online: https://www.kongsbergdiscovery.net/ek80/documents.htm (accessed on 15 May 2024).
48. MacLennan, D.N.; Simmonds, E.J. *Fisheries Acoustics*; Springer Science & Business Media: Berlin/Heidelberg, Germany, 2013; Volume 5, ISBN 94-017-1558-0.
49. Urick, R.J. *Principles of Underwater Sound*; McGraw-Hill Book Co.: New York, NY, USA, 1983.
50. Hart, A. Mann-Whitney Test Is Not Just a Test of Medians: Differences in Spread Can Be Important. *BMJ* **2001**, *323*, 391–393. [CrossRef] [PubMed]
51. Swathi Lekshmi, P.S.; Radhakrishnan, K.; Narayanakumar, R.; Vipinkumar, V.P.; Parappurathu, S.; Salim, S.S.; Johnson, B.; Pattnaik, P. Gender and Small-Scale Fisheries: Contribution to Livelihood and Local Economies. *Mar. Policy* **2021**, *136*, 104913. [CrossRef]
52. Pérez, N.P.; Guevara López, M.A.; Silva, A.; Ramos, I. Improving the Mann–Whitney Statistical Test for Feature Selection: An Approach in Breast Cancer Diagnosis on Mammography. *Artif. Intell. Med.* **2015**, *63*, 19–31. [CrossRef] [PubMed]
53. Sanyé-Mengual, E.; Specht, K.; Krikser, T.; Vanni, C.; Pennisi, G.; Orsini, F.; Gianquinto, G.P. Social Acceptance and Perceived Ecosystem Services of Urban Agriculture in Southern Europe: The Case of Bologna, Italy. *PLoS ONE* **2018**, *13*, e0200993. [CrossRef] [PubMed]
54. Nauen, J.C.; Lauder, G.V. Hydrodynamics of Caudal Fin Locomotion by Chub Mackerel, *Scomber japonicus* (*Scombridae*). *J. Exp. Biol.* **2002**, *205*, 1709–1724. [CrossRef] [PubMed]
55. Yang, Y.; Gastauer, S.; Proud, R.; Mangeni-Sande, R.; Everson, I.; Kayanda, R.J.; Brierley, A.S. Modelling and in Situ Observation of Broadband Acoustic Scattering from the Silver Cyprinid (*Rastrineobola argentea*) in Lake Victoria, East Africa. *ICES J. Mar. Sci.* **2023**, 1–14. [CrossRef]
56. Furusawa, M.; Miyanohana, Y. Application of Echo-Trace Analysis to Estimation of Behaviour and Target Strength of Fish. *Acoust. Sci. Technol.* **1988**, *9*, 169–180. [CrossRef]
57. Kanwisher, J.; Ebeling, A. Composition of the Swim-Bladder Gas in Bathypelagic Fishes. *Deep. Sea Res.* **1957**, *4*, 211–217. [CrossRef]
58. Benoit-Bird, K.J.; Au, W.W.; Kelley, C.D.; Taylor, C. Acoustic Backscattering by Deepwater Fish Measured in Situ from a Manned Submersible. *Deep. Sea Res. Part I Oceanogr. Res. Pap.* **2003**, *50*, 221–229. [CrossRef]
59. Albano, M.; D'iglio, C.; Spanò, N.; Fernandes, J.M.d.O.; Savoca, S.; Capillo, G. Distribution of the Order Lampriformes in the Mediterranean Sea with Notes on Their Biology, Morphology, and Taxonomy. *Biology* **2022**, *11*, 1534. [CrossRef] [PubMed]
60. Ladino, A.; Pérez-Arjona, I.; Espinosa, V.; Chillarón, M.; Vidal, V.; Godinho, L.M.; Boyra, G. Role of material properties in acoustical target strength: Insights from two species lacking a swimbladder. *Fish. Res.* **2024**, *270*, 106895. [CrossRef]
61. Rumape, O.; Elveny, M.; Suksatan, W.; Hatmi, R.U.; Voronkova, O.Y.; Bokov, D.O.; Wanita, Y.P. Study on the quality of fish products based on different preservation techniques: A review. *Food Sci. Technol.* **2022**, *42*, e78521. [CrossRef]
62. Yoon, E.; Oh, W.-S.; Lee, H.; Hwang, K.; Kim, D.-N.; Lee, K. Comparison of Target Strength of Pacific Herring (*Clupea pallasii* Valenciennes, 1847) from Ex-Situ Measurements and a Theoretical Model. *Water* **2021**, *13*, 3009. [CrossRef]
63. Sathish, K.; Anbazhagan, R.; Venkata, R.C.; Arena, F.; Pau, G. Investigation and Numerical Simulation of the Acoustic Target Strength of the Underwater Submarine Vehicle. *Inventions* **2022**, *7*, 111. [CrossRef]
64. Fässler, S.M.; O'Donnell, C.; Jech, J.M. Boarfish (*Capros aper*) Target Strength Modelled from Magnetic Resonance Imaging (MRI) Scans of Its Swimbladder. *ICES J. Mar. Sci.* **2013**, *70*, 1451–1459. [CrossRef]

65. Zhu, Y.; Mizutani, K.; Minami, K.; Shirakawa, H.; Kawauchi, Y.; Shao, H.; Tomiyasu, M.; Iwahara, Y.; Tamura, T.; Ogawa, M.; et al. Target Strength Measurements of Free-Swimming Sandeel Species, Ammodytes Spp., in a Large Indoor Experimental Aquarium. *J. Mar. Sci. Eng.* **2022**, *10*, 966. [CrossRef]
66. McClatchie, S.; Macaulay, G.; Coombs, R.F.; Grimes, P.; Hart, A. Target Strength of an Oily Deep-Water Fish, Orange Roughy (*Hoplostethus atlanticus*) I. Experiments. *JASA* **1999**, *106*, 131–142. [CrossRef]
67. Yan, N.; Mukai, T.; Hasegawa, K.; Yamamoto, J.; Fukuda, Y. Broadband target strength of arabesque greenling, Pacific sand lance, and pointhead flounder. *ICES J. Mar. Sci.* **2023**, *81*, 195–203. [CrossRef]
68. Lucca, B.M.; Warren, J.D. Experimental target strength measurements of pteropods and shrimp emphasize the importance of scattering model inputs. *ICES J. Mar. Sci.* **2024**, fsad211. [CrossRef]

fishes

MDPI

Article

A Fish Target Identification and Counting Method Based on DIDSON Sonar and YOLOv5 Model

Wei Shen [1,†], Mengqi Liu [1], Quanshui Lu [1], Zhaowei Yin [1] and Jin Zhang [1,2,3,*,†]

1 College of Oceanography and Ecological Science, Shanghai Ocean University, Shanghai 201306, China; wshen@shou.edu.cn (W.S.); m220200637@st.shou.edu.cn (M.L.); m210200566@st.shou.edu.cn (Q.L.); m230200712@st.shou.edu.cn (Z.Y.)

2 Office of Asset and Laboratory Management, Shanghai Ocean University, Shanghai 201306, China

3 College of Marine Living Resource Sciences and Management, Shanghai Ocean University, Shanghai 201306, China

* Correspondence: j_zhang@shou.edu.cn; Tel.: +86-15692161377

† These authrs contributed equally to this work.

Abstract: In order to more accurately and quickly identify and count underwater fish targets, and to address the issues of excessive reliance on manual processes and low processing efficiency in the identification and counting of fish targets using sonar data, a method based on DIDSON and YOLOv5 for fish target identification and counting is proposed. This study is based on YOLOv5, which trains a recognition model by identifying fish targets in each frame of DIDSON images and uses the DeepSort algorithm to track and count fish targets. Field data collection was conducted at Chenhang Reservoir in Shanghai, and this method was used to process and verify the results. The accuracy of random sampling was 83.56%, and the average accuracy of survey line detection was 84.28%. Compared with the traditional method of using Echoview to process sonar data, the YOLOv5 based method replaces the step that requires manual participation, significantly reducing the time required for data processing while maintaining the same accuracy, providing faster and more effective technical support for monitoring and managing fish populations.

Keywords: identification; counting; fish targets; YOLOv5 model; DeepSort algorithm

Key Contribution: This study combines YOLOv5 with DIDSON to develop a fast and accurate method for underwater fish target identification and counting. Compared to traditional methods, this approach significantly reduces data processing time while maintaining high accuracy, and it avoids the drawbacks of human fatigue and subjective bias.

1. Introduction

The identification and counting of fish targets are essential processes in the assessment and monitoring of fishery resources, playing a crucial role in sustainable management and conservation efforts. Accurate fish population estimates are vital for maintaining ecological balance, supporting commercial fisheries, and ensuring the long-term viability of aquatic ecosystems. Traditional studies of fish resources frequently employ acoustic techniques to circumvent the challenges posed by the underwater transmission of optical signals, notably in murky waters. However, these sonars, while providing an extensive detection range through echo integration or counting methods, are limited by insufficient recognition accuracy, which prevents them from fully meeting the demands of precise fish identification and counting [1].

The dual-frequency identification sonar (DIDSON), also known as the 'acoustic camera', delivers distinct acoustic images in obscured and dim underwater conditions [2] and has been widely used in fisheries management, underwater inspections, and environmental monitoring [3]. Research has demonstrated that DIDSON can effectively replace optical systems in murky waters, providing clear and nearly photographic images for various imaging tasks [4,5]. It has been employed for counting and measuring farmed fish during transfer [6], estimating fish abundance [7], and measuring swimming patterns and body length of cultured Chinese sturgeons [8].

Echoview, extensively utilized in hydroacoustic research, fisheries science, and marine environmental monitoring among other domains [9], proves effective for the processing and analysis of DIDSON data. Nevertheless, its semi-interactive, semi-automatic mode of data processing demands considerable time and incurs substantial labor costs when applied to the identification and counting of fish targets.

To address these limitations, researchers have explored alternative methods that aim to reduce manual intervention and increase processing efficiency. For instance, the use of custom MATLAB scripts has allowed for more tailored analysis, but these often require advanced programming skills and are not easily scalable. Recent developments in deep learning, particularly the application of Convolutional Neural Networks (CNNs) like Faster R-CNN, have shown potential in automating the detection process, though these approaches typically demand significant computational resources and may not be ideal for real-time applications. Hybrid methods that integrate traditional image processing with machine learning have also been proposed to enhance detection accuracy, but they tend to increase the complexity of processing, making them challenging to implement on a large scale.

Target recognition algorithms are fundamental in computer vision, encompassing both traditional machine learning and deep learning methods. With advancements in deep learning, deep learning-based target detection has increasingly become the preferred approach. YOLOv5, known for its rapid processing, high accuracy, and adaptability, has been widely employed across various applications, including acoustic image recognition. For example, YOLOv5 has been used to analyze audio data for animal species monitoring [10], identify single-fish echo trajectories on echo maps [11], recognize and localize targets in side-scan sonar images [12], and improve fish detection in noisy sonar environments to aid fish farming and resource assessment [13]. This demonstrates that YOLOv5 is highly effective and accurate in the field of acoustic image recognition.

This study combines YOLOv5 with DIDSON data to provide a new method for fish identification and counting, addressing the limitations of traditional methods like Echoview. By automating the detection process and eliminating the need for manual participation, YOLOv5 significantly reduces the time required for processing DIDSON data, avoiding the drawbacks of human fatigue and subjective bias, and consistently maintaining high accuracy. This method not only improves efficiency but also provides faster and more effective technical support for monitoring and managing fish populations, contributing to the sustainable development of fisheries.

2. Methodology

2.1. YOLOv5 Target Detection Model

YOLOv5, the fifth-generation model in the YOLO (You Only Look Once) series, was developed by Ultralytics for real-time object recognition tasks.

The architecture of YOLOv5 is segmented into four integral components: input, backbone, neck, and head [14]. These components work in concert to facilitate swift and precise target detection. The input layer is tasked with receiving and preprocessing image data to fulfill the requirements of the layers that follow. The backbone is primarily composed of multiple convolutional layers, normalization layers, and activation functions, dedicated to extracting features from images. The neck amplifies the capacity for feature representation using the feature pyramid network (FPN) and the path aggregation network (PAN). The head utilizes the extracted feature maps to determine the target's position, category, and confidence.

Additionally, YOLOv5 incorporates various efficient modules, such as spatial pyramid pooling (SPP) and the Focus module, which enhance the model's capability to detect objects across different scales and to extract intricate features. During training, a composite loss function is employed to refine various prediction tasks. At the inference stage, non-maximum suppression (NMS) is utilized to identify and retain the most accurate bounding box.

Pre-trained models of YOLOv5, available in various scales, can be chosen based on the specific needs of the task at hand. Furthermore, the model can be tailored and fine-tuned using custom datasets, broadening its applications in fields such as object recognition and motion target tracking.

2.2. DeepSort Multi-Object Tracking Algorithm

DeepSort is a multi-object tracking algorithm that leverages deep learning techniques. It utilizes an object detection model to process video frames sequentially, acquiring target positions and categories. Additionally, a deep learning model extracts high-dimensional feature vectors representing the appearance and motion characteristics of the targets. By employing the Kalman filter and the Hungarian algorithm, DeepSort matches and associates the detection results from the current frame with targets tracked in previous frames. This process determines the motion trajectory of the target, maintains its identity continuity, and computes the similarity between feature vectors to address challenges posed by target occlusion and resemblance in appearance. DeepSort constructs a trajectory model of the target using historical data and aligns this data based on the target's motion state and appearance, enhancing tracking precision. By integrating the appearance features, motion state, and historical data of the target, DeepSort facilitates efficient and accurate multi-target tracking in complex environments.

Following object detection on video frames by the YOLOv5 model, the DeepSort algorithm assigns each detected object a bounding box, category label, and confidence level. An additional feature extraction network extracts appearance features, aiding in distinguishing different objects during the tracking phase.

$$d(i,\ j) = \sqrt{(Z_i - H\hat{x}_j)S_j^{-1}(Z_i - H\hat{x}_j)^T} \tag{1}$$

The algorithm applies a Kalman filter to forecast each tracked object's position, producing a predicted state. It then uses the Mahalanobis distance $d(i,j)$ to evaluate the discrepancy between the Kalman predicted state and the newly detected object, incorporating covariance considerations to optimize the matching process.

In the DeepSort algorithm, the Mahalanobis distance quantifies the discrepancy between the ith detection and the jth prediction. This Mahalanobis distance measures the difference between the newly detected observation Z_i, and the predicted state $H\hat{x}_j$, factoring in the covariance matrix S_j of the measurement prediction.

The Mahalanobis distance effectively explains the correlation between variables and provides a more accurate measure of the similarity between predicted states and the newly detected object. This ability makes it particularly suitable for improving the accuracy of object matching in the DeepSort algorithm.

$$c(i,j) = 1 - \frac{f_i \cdot f_j}{||f_i||\,||f_j||} \tag{2}$$

Furthermore, the cosine distance is utilized to assess the similarity between the appearance features of detected and tracked objects, facilitating the differentiation of individual targets.

This is denoted as $c(i,j)$, where f_i and f_j are the appearance feature vectors of the detected and tracked objects, respectively. The value range of the formula is between [0,2], where a smaller value indicates that the two vectors are more similar and conversely, a larger value indicates they are less similar.

The cosine distance effectively measures the directional difference between two feature vectors, while ignoring their magnitude. This property is particularly useful when tracking applications that have different object sizes but still have similar appearance characteristics. The smaller the cosine distance, the higher the similarity, which helps to accurately match targets with similar appearances.

DeepSort incorporates a cascade matching strategy, which prioritizes the matching of long-standing trajectories to enhance tracking consistency and precision. IoU matching serves as a supplementary method to further support tracking stability. When the tracked object's position and features are updated based on the matching results, its continuity is maintained effectively. However, if a tracked object fails to match any detection across consecutive frames, it is deemed to have disappeared, and the tracking process for that object is consequently terminated.The specific flowchart of the entire algorithm is shown in Figure 1.

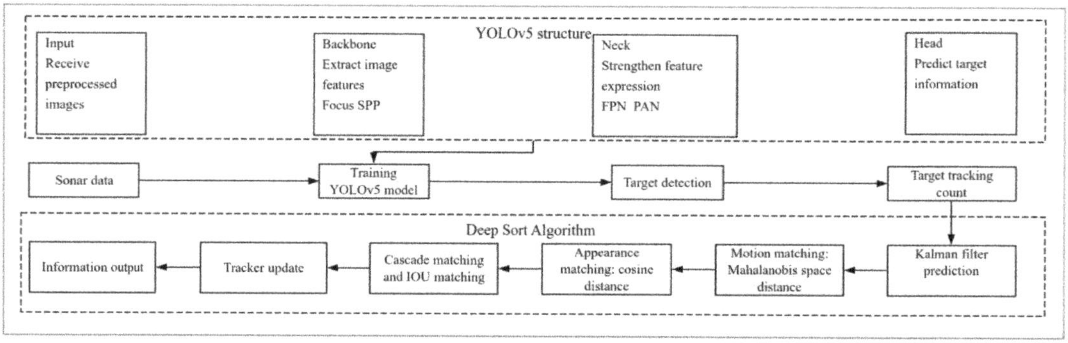

Figure 1. Algorithm flowchart. The figure illustrates the integrated workflow of YOLOv5 and DeepSORT for object tracking. Initially, YOLOv5 extracts image features from the input sonar data, completing object detection through its Backbone, Neck, and Head modules. Following detection, the DeepSORT algorithm takes over, using Kalman filtering for prediction, Mahalanobis distance for motion matching, and cosine distance for appearance matching. Finally, the algorithm updates the target trajectories through cascade matching and IoU matching, outputting the tracking results. This workflow effectively combines detection and tracking, ensuring accurate multi-object tracking even in complex environments.

2.3. Field Data Acquisition

2.3.1. DIDSON Dual-Frequency Identification Sonar

DIDSON, a high-resolution identification sonar created by the University of Washington and produced by Sound Metrics in the United States, utilizes the sound wave focusing principle of sound lenses to form a narrow beam. This technology enables the production of images nearly equivalent to optical quality in low-visibility underwater conditions, as illustrated in Figure 2. The acoustic lens, which requires minimal power for beam compression, facilitates the transmission and reception of the beam. Such a configuration enhances operational efficiency and decreases the size of the equipment [15]. The sonar operates with a horizontal angle of 29° and a vertical angle of 14° [16]. It is capable of delivering high-resolution images at 0.3° with a frequency of 1.8 MHz, providing distinctly clear images up to a distance of 11.63 m. At a reduced frequency of 1.1 MHz, the sonar achieves a resolution of 0.6° and a maximum detection range of 40 m, where it can automatically focus on targets and maintain image clarity within a 1–40 m range. The specific parameters of DIDSON are shown in Table 1.

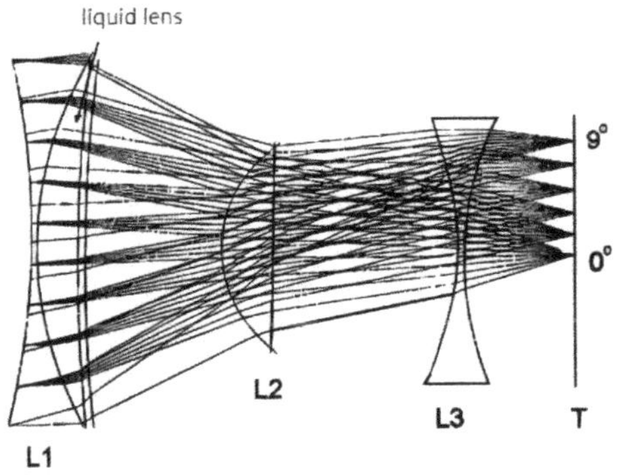

Figure 2. DIDSON imaging schematic. L1 is a lens triplet composed of a biconcave plastic lens, a liquid medium, and a thinner plastic lens. L2 is a plano-convex plastic lens, and its focal length can be adjusted by altering the distance between it and L1. L3 is positioned in front of the transducer array T. When sound waves are incident on L1 at a 0° angle, the focal point of the waves aligns with the center of the transducer array. When sound waves are incident at a 9° angle, the lens alters the propagation path of the waves, focusing them at the 9° position of the transducer array.

Table 1. DIDSON sonar related parameters.

Specification/Mode	Low Frequency	High Frequency
Operating Frequency	1.0 MHz	1.8 MHz
Beam Width	Horizontal 0.4°, Vertical 12°	Horizontal 0.3°, Vertical 12°
Number of Beams	48	96
Source Level	202 dB re 1 μPa at 1 m	206 dB re 1 μPa at 1 m
Start Range	0.75 m to 40 m	0.38 m to 11.63 m
Maximum Frame Rate	4–21 frames/s	
Field of View	29°	
Remote Focusing	1 m to maximum range	
Power Consumption	Watts typical	
Weight in Air	7.0 kg (15.4 lbs)	
Weight in Water	−0.61 kg (1.33 lbs)	
Dimensions	30.7 cm × 20.6 cm × 17.1 cm	

2.3.2. Data Acquisition

In October 2023, a motorized boat equipped with DIDSON was utilized for a survey of fish resources in Chenhang Reservoir, Shanghai, China. The sonar was mounted on the starboard side at a draft of 0.5 m and angled downward at 60°. The direction of data collection aligned with the forward motion of the boat. Custom brackets were implemented to minimize vibrations of the device during movement, thereby ensuring the capture of high-quality images. The survey's path is depicted in Figure 3. Considering the conditions of the Chenhang Reservoir, the high-frequency mode was used throughout. The window start was set at 0.83 m, with a window length of 11.63 m. The sampling rate was 8 frames per second, the receiving gain was set to 25, and the threshold was around 15, adjusted according to the actual condition.

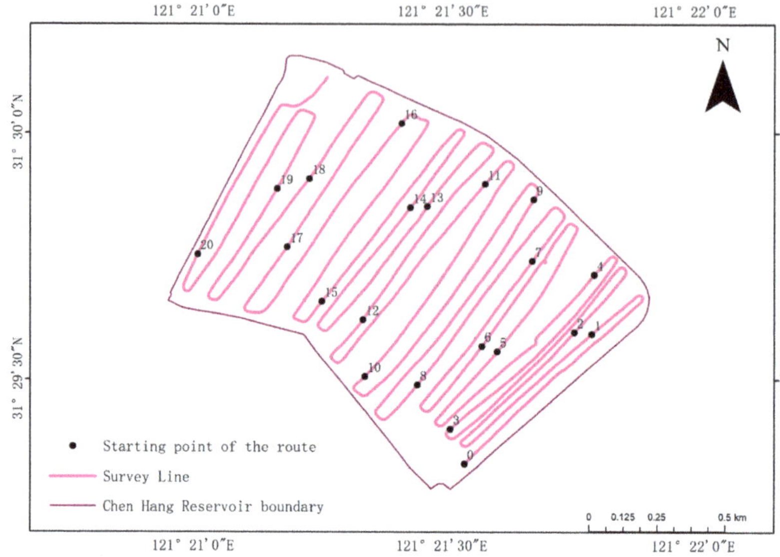

Figure 3. Chenhang Reservoir sonar survey route. During the field sonar data collection process, starting from the bottom right corner of the image and ending at the top left corner, we switch the survey line file approximately every eight minutes to ensure consistent data file sizes, with a measurement distance of approximately 1 km, making subsequent processing more convenient.

2.4. YOLOv5 Model Training

Among the 88,763 frames of images collected by DIDSON, 8876 were randomly chosen, and 1000 containing fish targets were identified. Out of these, 100 images were designated for the validation set, and 900 were used for the training set.

The image annotation tool Labellmg was employed in this study. Initially, the selected images were uploaded to Labellmg, where the fish targets were manually annotated, as depicted in Figure 4. This process generated an XML file for each annotated image, detailing the position coordinates and category labels of the fish bounding boxes. These XML files were formatted according to the PASCAL VOC standards for further model training.

All raw images and XML files were stored in the designated directory for the YOLOv5 training model. Modifications were made to the target category, which was set to 1 and labeled "fish". Adjustments to the training parameters included setting the epochs to 300, the batch size to 8, the number of workers to 4, and the image size to 640 × 640.

Upon completion of the training, all training data were saved in the results file, and a line graph was generated to display the precision (P), recall (R), loss, average precision (AP), as shown in Figure 5.

$$P = \frac{TP}{TP + FP} \tag{3}$$

$$R = \frac{TP}{TP + FN} \tag{4}$$

$$AP = \int_0^1 P(R)dR \tag{5}$$

In this graph, TP denotes the number of correctly identified fish targets, FP indicates the falsely identified fish targets, and FN represents the fish targets that were not detected.

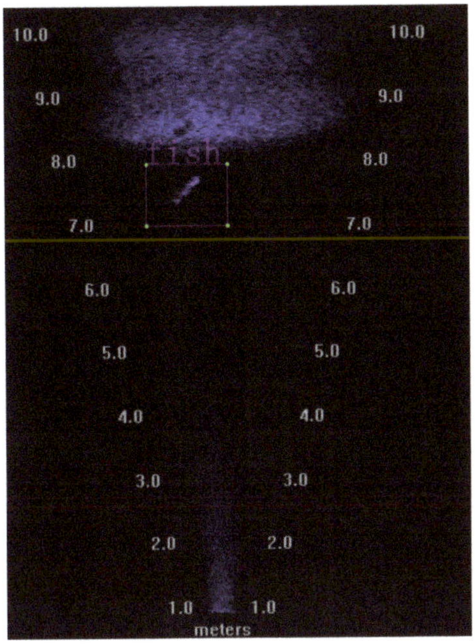

Figure 4. Labellmg annotation schematic. In this figure, the area from 1 to 4 m shows speckle noise caused by factors such as ship noise, bubbles, and impurities in the water. The boxed section highlights a horizontally oriented fish. Due to the relatively high speed of the vessel, the sonar image of the fish appears as connected blocks, creating a jagged or sawtooth pattern. The large reflective area from 8 to 10 m is caused by reflections from the lakebed.

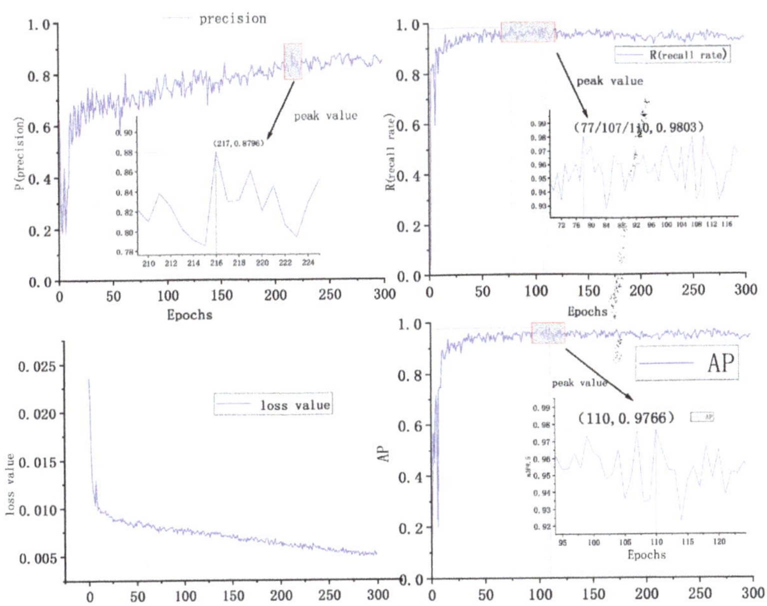

Figure 5. Model evaluation parameters.

According to the data in Figure 5, it was observed that, as the number of iterations increases, the accuracy in the top left stabilizes at around 0.85, reaching a peak of 0.8796 at 217 iterations. The recall in the top right stabilizes at around 0.95, reaching its maximum value of 0.9803 at 77, 107, and 110 epochs. The loss value in the bottom left gradually decreases as the model converges. The AP (average precision) in the bottom right stabilizes at around 0.95, with a peak value of 0.9766 achieved at 110 epochs. These evaluation results suggest that, after 300 iterations, the model achieved relative stability and performed effectively in recognizing fish targets in image data.

3. Experiments and Analyses

3.1. Target Identification and Counting

Upon completion of the YOLOv5 model training, configure the DeepSort parameters (such as the maximum tracking count of 70 frames, a minimum detection count of 3 frames, a minimum confidence threshold of 0.3, a maximum cosine distance of 0.2, a maximum IOU distance of 0.7, and a maximum overlap ratio for non-maximum suppression (NMS) of 1.0,etc.) Then, the model was integrated with the sonar video files requiring analysis, and the YOLOv5 model will pass the detection results to DeepSort for tracking. By assigning a unique ID to each detected fish target and tracking its movement, accurate and rapid fish target recognition and counting are achieved, as depicted in Figure 6. This survey consisted of 21 lines, with the processed results summarized in Table 2.The 'Survey Line' row represents the survey line number, which can be seen in Figure 3, while the 'YOLOv5' row indicates the number of fish detected on that specific survey line, with a total of 1760 fish identified.

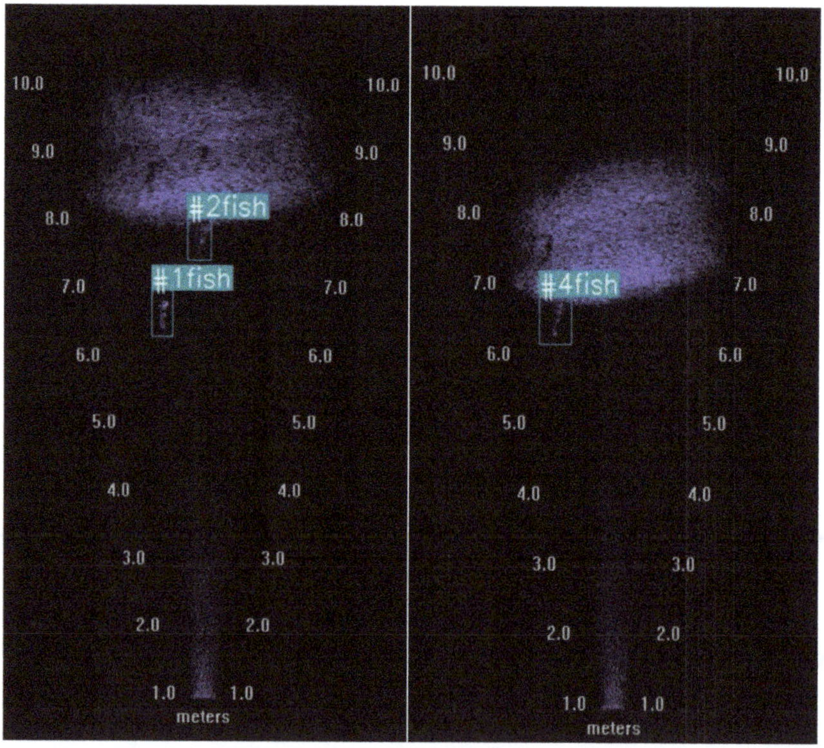

Figure 6. Fish target identification.

Table 2. Automatic identification count.

Survey Line	0	1	2	3	4	5	6	7	8	9	10
YOLOv5 (fish)	188	216	143	122	145	100	70	49	71	100	63
Survey Line	11	12	13	14	15	16	17	18	19	20	Total
YOLOv5 (fish)	70	66	12	44	36	48	134	44	22	17	1760

Note: The values represent the number of fish automatically identified by YOLOv5 on each survey line.

Table 3 presents the output data of the identified fish targets, "ID" represents the identification number for each fish. "Frame Count" indicates the total number of frames in which the fish target was detected. "First Frame" represents the frame number where the tracked object first appeared, and "Last Frame" indicates the frame number where the tracked object last appeared. "Avg_Depth" represents the actual average depth of the fish after data conversion. "Lat" and "Lon" represent the latitude and longitude where the tracked object appeared. This information is instrumental for subsequent analyses concerning the distribution of fish density.

Table 3. Information related to fish target identification.

ID	Frame Count	First Frame	Last Frame	Avg_Depth (m)	Lat	Lon
1	11	4	14	5.87	31.49176567	121.355733
2	13	7	19	6.70	31.49177917	121.3557462
3	14	21	34	6.36	31.4918065	121.3557723
4	7	121	127	6.43	31.49198483	121.355938
5	4	155	158	5.40	31.49204017	121.355989

3.2. Accuracy Evaluation

3.2.1. Random Sampling

From the datasets of the initial five survey lines, a tenth of the images were randomly selected. An accuracy analysis was conducted on 500 images containing fish targets, with the results compared against manual recognition outcomes, as shown in Table 4. The identification accuracy reached 83.56%, with the primary causes of recognition errors listed as follows:

1. Excessive clustering of fish led to errors in identification.
2. Errors were likely when fish targets neared the water bottom due to linear propagation, causing them to resemble the bottom.
3. Complex underwater terrain contributed to recognition errors.

Table 4. Evaluation of the accuracy of extracted images.

	Manual Identification	Total Identifications	Correct Identifications	Unidentified	Misidentifications	Accuracy (%)	Unidentification Rate (%)	Misidentification Rate (%)
Count (fish)	675	588	564	87	24	83.56%	12.89%	3.56%

Note: The values represent the number of fish.

3.2.2. Line Inspection

Data from three survey lines were selected to conduct statistics on fish targets and evaluate accuracy, which were then compared with manual recognition results, as presented in Table 5. The average accuracy rate was 84.28%, marginally higher than the sampling rate. This increase was attributed to the movement of both the fish and the boat, causing the same fish targets to appear in multiple frames and to be recognized and counted repeatedly.

Table 5. Evaluation of line accuracy.

Survey Line	Manual identification Count	YOLOv5 Total Identifications Count	Correct Identifications Count	Unidentified Count	Misidentifications Count	Accuracy (%)	Unidentification Rate (%)	Misidentification Rate (%)
10	71	63	61	8	2	85.92%	11.27%	2.82%
11	83	70	69	13	1	83.13%	15.66%	1.20%
12	75	66	63	9	3	84%	12%	4%
Average				10	3	84.28%	13.10%	2.62%

Note: The values represent the number of fish automatically identified by YOLOv5 on each survey line.

Using the DIDSON data from these three survey lines, a comparison was made with different versions of the YOLO method and the traditional Echoview method for processing DIDSON data. The results are shown in Table 6. It can be seen that the accuracy of several methods on the three survey lines is not significantly different but in terms of processing speed, the Echoview method requires a lot of time and manual involvement throughout the entire process. When collecting data from deeper waters, DIDSON sonar data will experience more noise and interference, and the time and effort required for manual work will double. Therefore, the method combining YOLOv5 and DIDSON has strong practicality, saving a lot of the time and energy required for manual processing.

Table 6. Comparison of Accuracy Across Different Methods.

Method	YOLOv5	YOLOv6	YOLOv8	Echoview	Manual Identification
Survey Line 10	63	61	59	60	71
Survey Line 11	70	63	64	69	83
Survey Line 12	66	67	75	70	75
Processing Time (single survey line)	3 min	3 min	3 min	Approximately 30 min	Approximately 120 min
Deviation (total)	30	38	31	30	

Note: These values represent the number of fish identified by different methods.

4. Discussion

4.1. Sonar Images

As a high-resolution recognition sonar, DIDSON excels in identifying and counting fish targets underwater compared to traditional scientific fish detectors. However, several factors still impede its accuracy enhancement. For example, sonar echoes reflected by bubbles, plankton, tree branches, and debris in water bodies can introduce noise into DIDSON sonar images [17]. Environmental noise during the collection process, such as ship engine sounds and noise generated by waves, can also affect the sonar images, complicating the detection of fish targets [18]. Balk et al. [19] noted that high boat speeds could result in jagged contours of collected fish targets, a problem that could be alleviated by reducing the boat's speed. Furthermore, the behavior of fish targets, such as clustering or nearing the bottom, could cause targets to overlap [20] or merge, leading to inaccuracies in their counts. Therefore, the following measures should be taken to obtain high-quality sonar images:

1. Maintaining an appropriate vessel speed, ideally around 5–6 km/h
2. Choose a vessel with minimal noise.
3. Ensure the equipment is securely installed to avoid vibrations.
4. Whenever possible, select clean water areas to reduce reflections from debris in the water.

4.2. Identification and Counting

YOLOv5 combined with DeepSort offers substantial benefits in the practical applications of fish target identification and counting. YOLOv5, known for its high precision and rapid response, is well-suited for real-time object identification and counting [21]. Optimal identification performance necessitates extensive training and validation with a significant amount of data, which should be diverse and cover various scenarios, including different fish types, postures, and densities, as well as a range of underwater conditions and lighting situations [22]. Moreover, precise adjustment of algorithm parameters is crucial; the selection of weight parameters in YOLOv5 model training greatly influences model performance. Adjustments must be made to different loss terms in the loss function, such as classification loss, localization loss, and confidence loss, to meet specific application requirements. An adequate number of training epochs is essential, as too few can result in under-learning, while too many might lead to overfitting and reduce the model's ability to generalize to new data [23]. DeepSort plays a key role in continuously tracking and counting targets in video frames [24]. It utilizes Kalman filters to predict target positions and employs deep learning features alongside the Hungarian algorithm to resolve target association challenges [23]. DeepSort leverages pre-trained convolutional neural networks to extract appearance features of targets, encoding these into fixed-length vectors to assess target similarity. This process effectively manages challenges posed by temporarily occluded or densely clustered fish targets, enhancing the robustness and reliability of tracking and counting. However, DeepSort demands significant computational resources, such as GPU and memory, particularly when handling high-resolution videos and numerous targets. The performance of DeepSort is heavily dependent on the quality of the feature extraction network; inadequacies in this network can lead to incorrect target associations. In scenarios involving highly clustered or swiftly moving targets, DeepSort might encounter tracking inaccuracies or losses, a limitation that can be partially addressed through deep learning features but not entirely eliminated. By strategically allocating computing resources, optimizing feature extraction networks, and thoroughly preparing training data, the combined capabilities of YOLOv5 and DeepSort can be effectively utilized to enhance fish target recognition and tracking performance.

4.3. Manual Identification

During the accuracy evaluation and verification phases, manual counting was employed for validation. The results varied among individuals due to differences in their levels of cognition and experience in recognizing fish targets [25], precluding the possibility of achieving absolute accuracy. Furthermore, prolonged periods of manual counting were susceptible to errors stemming from human fatigue. Although YOLOv5 demonstrated a slightly lower accuracy compared to manual counting, it provided the advantage of stable, automatic counting, which is not affected by fatigue or subjective biases. This feature is particularly beneficial for the consistent identification and counting of fish targets.

5. Conclusions

The importance of automated identification and counting of fish targets cannot be overstated in the context of fish resource assessment and management. This study introduced a method that integrates high-definition DIDSON sonar with YOLOv5, achieving a recognition accuracy exceeding 80%. This method offers substantial improvements over traditional Echoview method in terms of recognition accuracy, processing speed, and reduced labor costs.

The accuracy of the proposed method is influenced by several factors, including the quality of sonar images, background noise, and variations in fish size, density, and proximity to the water bottom. Further refinement is necessary to enhance identification and counting accuracy while minimizing disturbances such as acoustic noise during data acquisition. Additionally, the parameter settings and data training for YOLOv5 and DeepSort require optimization to support the model's generalization capabilities and robustness across various application scenarios. Future research will focus on enhancing the accuracy and reliability of fish target identification and counting by improving sonar technology, refining model structures, and enriching the diversity of training data.

In summary, the method proposed in this study offers high recognition accuracy and significantly improves the speed of sonar data processing, while avoiding the drawbacks of human fatigue and subjective bias. It consistently maintains a high level of accuracy, enhances efficiency, and provides faster and more effective technical support for the monitoring and management of fish populations, thereby contributing to the sustainable development of fisheries.

Author Contributions: W.S., Conceptualization, formal analysis, investigation, supervision and writing—review and editing; M.L., investigation, methodology, writing—original draft preparation and, writing—review and editing and data curation. Q.L., validation, writing—original draft preparation and resources. Z.Y., data curation, resources and writing—review and editing. J.Z., investigation, supervision, project administration, resources and formal analysis. All authors have read and agreed to the published version of the manuscript.

Funding: This research received no external funding.

Institutional Review Board Statement: Not applicable.

Informed Consent Statement: Not applicable.

Data Availability Statement: All data used in this study can be obtained from the corresponding author.

Acknowledgments: The authors sincerely appreciate all the reviewers for their invaluable feedback and suggestions.

Conflicts of Interest: The authors declare no conflicts of interest.

References

1. Cronkite, G.M.W.; Enzenhofer, H.J. Observations of Controlled Moving Targets with Split-Beam Sonar and Implications for Detection of Migrating Adult Salmon in Rivers. *Aquat. Living Resour.* **2002**, *15*, 1–11. [CrossRef]
2. Perivolioti, T.-M.; Tušer, M.; Terzopoulos, D.; Sgardelis, S.P.; Antoniou, I. Optimising the Workflow for Fish Detection in DIDSON (Dual-Frequency IDentification SONar) Data with the Use of Optical Flow and a Genetic Algorithm. *Water* **2021**, *13*, 1304. [CrossRef]
3. Moursund, R.A.; Carlson, T.J.; Peters, R.D. A Fisheries Application of a Dual-Frequency Identification Sonar Acoustic Camera. *ICES J. Mar. Sci.* **2003**, *60*, 678–683. [CrossRef]
4. Belcher, E.; Hanot, W.; Burch, J. Dual-Frequency Identification Sonar (DIDSON). In Proceedings of the 2002 Interntional Symposium on Underwater Technology (Cat. No.02EX556), Tokyo, Japan, 19 April 2002; pp. 187–192.
5. Maxwell, S.L.; Gove, N.E. *The Feasibility of Estimating Migrating Salmon Passage Rates in Turbid Rivers Using a Dual Frequency Identification Sonar (DIDSON)*; Anchorage, Alaska Department of Fish and Game, Division of Commercial Fisheries: Anchorage, Alaska, 2004.
6. Han, J.; Honda, N.; Asada, A.; Shibata, K. Automated Acoustic Method for Counting and Sizing Farmed Fish during Transfer Using DIDSON. *Fish. Sci.* **2009**, *75*, 1359–1367. [CrossRef]
7. Jing, D.; Han, J.; Wang, X.; Wang, G.; Tong, J.; Shen, W.; Zhang, J. A Method to Estimate the Abundance of Fish Based on Dual-Frequency Identification Sonar (DIDSON) Imaging. *Fish. Sci.* **2017**, *83*, 685–697. [CrossRef]
8. Zhang, H.; Wei, Q.; Kang, M. Measurement of Swimming Pattern and Body Length of Cultured Chinese Sturgeon by Use of Imaging Sonar. *Aquaculture* **2014**, *434*, 184–187. [CrossRef]
9. Ladroit, Y.; Escobar-Flores, P.C.; Schimel, A.C.G.; O'Driscoll, R.L. ESP3: An Open-Source Software for the Quantitative Processing of Hydro-Acoustic Data. *SoftwareX* **2020**, *12*, 100581. [CrossRef]
10. Husain, B.H.; Osawa, T. Advancing Fauna Conservation through Machine Learning-Based Spectrogram Recognition: A Study on Object Detection Using YOLOv5. *J. Sumberd. Alam Dan Lingkung.* **2023**, *10*, 58–68. [CrossRef]
11. Tong, J.; Wang, W.; Xue, M.; Zhu, Z.; Han, J.; Tian, S. Automatic Single Fish Detection with a Commercial Echosounder Using YOLO v5 and Its Application for Echosounder Calibration. *Front. Mar. Sci.* **2023**, *10*, 1162064. [CrossRef]
12. Yu, Y.; Zhao, J.; Gong, Q.; Huang, C.; Zheng, G.; Ma, J. Real-Time Underwater Maritime Object Detection in Side-Scan Sonar Images Based on Transformer-YOLOv5. *Remote Sens.* **2021**, *13*, 3555. [CrossRef]
13. Xing, B.; Sun, M.; Ding, M.; Han, C. Fish Sonar Image Recognition Algorithm Based on Improved YOLOv5. *Math. Biosci. Eng.* **2024**, *21*, 1321–1341. [CrossRef] [PubMed]
14. Li, Z.; Song, J.; Qiao, K.; Li, C.; Zhang, Y.; Li, Z. Research on Efficient Feature Extraction: Improving YOLOv5 Backbone for Facial Expression Detection in Live Streaming Scenes. *Front. Comput. Neurosci.* **2022**, *16*, 980063. [CrossRef] [PubMed]
15. Belcher, E.O.; Lynn, D.C.; Dinh, H.Q.; Laughlin, T.J. Beamforming and Imaging with Acoustic Lenses in Small, High-Frequency Sonars. In Proceedings of the Oceans '99. MTS/IEEE. Riding the Crest into the 21st Century. Conference and Exhibition. Conference Proceedings (IEEE Cat. No.99CH37008), Seattle, WA, USA, 13–16 September 1999; Volume 3, pp. 1495–1499.
16. Holmes, J.A.; Cronkite, G.M.W.; Enzenhofer, H.J.; Mulligan, T.J. Accuracy and Precision of Fish-Count Data from a "Dual-Frequency Identification Sonar" (DIDSON) Imaging System. *ICES J. Mar. Sci.* **2006**, *63*, 543–555. [CrossRef]

17. Kovesi, P. Phase Preserving Denoising of Images. *Signal* **1999**, *4*, 212–217.
18. Luo, Y.; Lu, H.; Zhou, X.; Yuan, Y.; Qi, H.; Li, B.; Liu, Z. Lightweight Model for Fish Recognition Based on YOLOV5-MobilenetV3 and Sonar Images. *Guangdong Nongye Kexue* **2023**, *50*, 37–46.
19. Balk, H.; Lindem, T. Improved Fish Detection in Data from Split-Beam Sonar. *Aquat. Living Resour.* **2000**, *13*, 297–303. [CrossRef]
20. Helminen, J.; Linnansaari, T. Object and Behavior Differentiation for Improved Automated Counts of Migrating River Fish Using Imaging Sonar Data. *Fish. Res.* **2021**, *237*, 105883. [CrossRef]
21. Redmon, J.; Farhadi, A. YOLOv3: An Incremental Improvement. *arXiv* **2018**, arXiv:1804.02767.
22. Lin, T.-Y.; Maire, M.; Belongie, S.; Hays, J.; Perona, P.; Ramanan, D.; Dollár, P.; Zitnick, C.L. Microsoft COCO: Common Objects in Context. In Proceedings of the Computer Vision—ECCV 2014; Fleet, D., Pajdla, T., Schiele, B., Tuytelaars, T., Eds.; Springer International Publishing: Cham, Switzerland, 2014; pp. 740–755.
23. Shafiq, M.; Gu, Z. Deep Residual Learning for Image Recognition: A Survey. *Appl. Sci.* **2022**, *12*, 8972. [CrossRef]
24. Wojke, N.; Bewley, A.; Paulus, D. Simple Online and Realtime Tracking with a Deep Association Metric. In Proceedings of the 2017 IEEE International Conference on Image Processing (ICIP), Beijing, China, 17–20 September 2017; pp. 3645–3649.
25. Keefer, M.L.; Caudill, C.C.; Johnson, E.L.; Clabough, T.S.; Boggs, C.T.; Johnson, P.N.; Nagy, W.T. Inter-Observer Bias in Fish Classification and Enumeration Using Dual-Frequency Identification Sonar (DIDSON): A Pacific Lamprey Case Study. *Northwest Sci.* **2017**, *91*, 41–53. [CrossRef]

fishes

MDPI

Review

Application of Fisheries Acoustics: A Review of the Current State in Mexico and Future Perspectives

Violeta E. González-Máynez [1,2], Enrique Morales-Bojórquez [3,*], Manuel O. Nevárez-Martínez [2] and Héctor Villalobos [4]

1 Centro de Investigaciones Biológicas del Noroeste, Guaymas 85400, Sonora, Mexico; vmaynez@pg.cibnor.mx
2 Instituto Mexicano de Investigación en Pesca y Acuacultura Sustentable, Guaymas 85400, Sonora, Mexico; manuel.nevarez@imipas.gob.mx
3 Centro de Investigaciones Biológicas del Noroeste, La Paz 23096, Baja California Sur, Mexico
4 Instituto Politécnico Nacional, CICIMAR, La Paz 23096, Baja California Sur, Mexico; hvillalo@ipn.mx
* Correspondence: emorales@cibnor.mx

Abstract: In Mexico, marine acoustics research still faces technical and scientific challenges. For the past decade, the country has made a sustained effort to implement acoustic techniques to generate time series of standardized information; however, these data have been underutilized. Marine acoustics research has been used mainly for small pelagic species and has contributed to improving fishery management and to advising stakeholders. The Mexican scientific community has perceived marine acoustic techniques as expensive tools that are only used for industrial fishing purposes. Marine acoustics can provide information on the variability and interactions between species, their physical environment, and other communities of species, but this approach has not yet been integrated into interdisciplinary research programs or ecosystem models. Additionally, acoustic data provide estimates of biomass and indices of relative abundance, and they have suitable statistical properties for use in integrated catch-at-age models. In summary, to consolidate marine acoustic techniques in Mexico, it is necessary, at a minimum, to maintain the current infrastructure for acoustic studies, to increase the budget for the development of monitoring programs that collect ecosystem indicator data, to promote the training of human resources, and to encourage peer review of the information generated and reported in gray literature.

Keywords: small pelagic; acoustic techniques; index of relative abundance

Key Contribution: (1) At present, it is critical for Mexico to consolidate the use of acoustic techniques and to produce added value for the acoustic data that are currently generated. Hence, a discussion on the contributions of acoustics to fishery management, Marine Protected Areas, and the ecosystem approach among the community of Mexican fisheries scientists should be promoted. (2) Investment in scientific instrumentation and human capital is relatively expensive in the short term, but fisheries acoustics has the potential to solve many of the challenges facing Mexican fisheries.

1. Introduction

Scientific echosounders have positioned themselves as extensively used tools in abundance and biomass assessments of resources as diverse as mesopelagic fishes, pelagic fishes, semidemersal organisms, macroalgal forests, zooplankton, squid, jellyfish, and top predators [1–9]. As technologies and data analysis methods have matured, the use of acoustics in ecological studies has grown rapidly [10,11], even developing in benthic habitat mapping to classify substrates and habitat type [12].

The constant and progressive evolution of acoustic technology makes tracing the historical events that led to the application of acoustics in fisheries research a difficult task. At the beginning of the First World War, there was a great push for research on transducers due to their military applications; two decades after the war, echosounders had become widespread, and acoustics were already being used as an assessment tool in European, American, Asian, and African countries, as well as in Australia and New Zealand [13–16]. In Mexico, during the 1970s, researchers from the Mexican Institute for Research in Sustainable Fisheries and Aquaculture (in Spanish: Instituto Mexicano de Investigación en Pesca y Acuacultura Sustentables (IMIPAS)) carried out acoustic surveys using analog echosounders off the coast of Baja California to estimate the anchovy biomass [17]. However, there was a failure to consolidate the use of these techniques in the country, and by the end of 1980, acoustic techniques for the evaluation of anchovy populations in Mexico had been replaced by the daily egg production method (DEPM) [18].

During the next two decades (1980–2000), Mexico did not make significant investments in acoustic equipment, nor in the training of human resources to develop and promote this line of research in the country. In 2007, the Mexican government, through IMIPAS and in collaboration with researchers from the French Institute for Ocean Science (IFREMER), began a fisheries acoustics monitoring program in the Gulf of California for small pelagic species using a single-beam 38 kHz frequency echosounder. Another frequency, 120 kHz, was added in 2012. Subsequently, new acoustic monitoring programs were implemented in different regions such as the western coast of the Baja California Peninsula, the northern region of Nayarit, and off the coast of Sinaloa (Figure 1). The main reason and rationale for studying the small pelagic fishery is that sardines and anchovies caught in northwestern Mexico represent the greatest contribution to the national fishery catch [19]. To date, acoustic techniques have been used most frequently to study these resources.

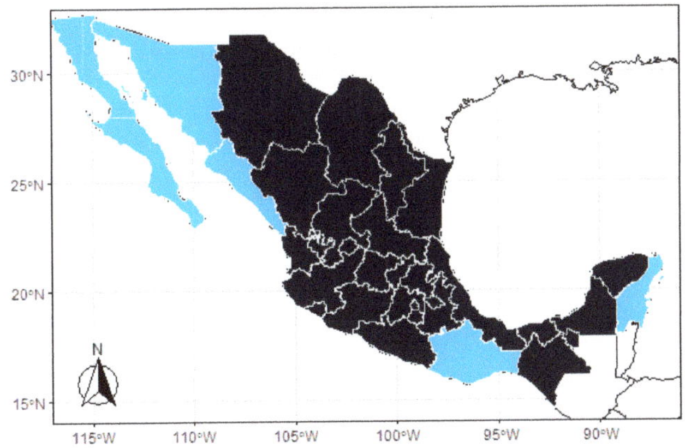

Figure 1. The Mexican states in which acoustics studies have been applied to Mexican commercial fisheries are denoted in light blue. These studies have been developed almost entirely in northwestern Mexico, with limited studies in Oaxaca and Quintana Roo.

Since 2012, there have been several publications in which acoustics have played a central role in stock assessment and fishery management in Mexico. However, these investigations have mainly focused on the northwestern Pacific coast of Mexico, including the Gulf of California. The objective of this review is to analyze and discuss the applications of acoustic techniques in Mexico, including their successes, utility, and limitations compared with those in other countries. We hope that this review will encourage the development and application of these techniques in Mexico for a wide range of ecological topics outside of the standard procedures for monitoring Mexican marine resources.

2. Materials and Methods

This review was based on manuscripts with the following eligibility criteria: (i) published and available from Scopus Elsevier, Redalyc, Crossref Metadata, and Google Scholar;

(ii) research papers or technical reports mainly from Institute of the Sea of Peru (IMARPE), IFREMER, the National Oceanic and Atmospheric Administration (NOAA), the Commonwealth Scientific and Industrial Research Organization (CSIRO), and IMIPAS; and (iii) studies with an emphasis on Mexican marine resources. Gray literature on acoustic research was excluded (e.g., abstracts in symposiums or congresses). The review is organized into six topics, namely processing software, sampling design, frontiers for the applications of acoustics, the relevance of time series, the Mexican budget for research, and conclusions.

3. Processing Software

Acoustic work requires real-time data acquisition and processing software. There are a limited number of manufacturers and suppliers of scientific echosounders, typically Biosonics or Konsgberg, with the choice of supplier depending on the specific requirements and applications. In Mexico, the echosounders are SIMRAD EK60 and EK80, both provided by Konsgberg (Horten, Norway). Comparatively, the choice of processing software is not simple. A research group can choose to pay for a license or use freely available software, but the available options differ greatly in price, development level, and available guidance, all of which influence the analysis of the available information and its reproducibility. The processing sequence sometimes involves manual processing providing physical quantities (e.g., backscatter) that are used to compute the density and biomass values for a fishery resource. The software shares some specific characteristics, such as an array of colors, or "palette", that are mapped to the backscatter; algorithms for the correction of a line that describes the bottom, which discard possible contributions from the seabed in the water column; algorithms for removing noise and reverberation; and other attributes. This review is based on the most used and frequently cited software in the literature (Table 1).

Echoview (ver. 8-15) has been cited the most in scientific publications. It was developed by an Australian company of the same name (formerly known as Myriax) via collaboration and feedback from the research community. It is used and distributed in several countries. Access to the program requires a physical key that can be shared within the workgroup (without being used at the same time), while workflows can be made available to other users who have access to the same or later versions of the program in which they were made. Payment for the annual renewal of the license allows the user to access the latest updates and innovative methodologies, while the modules vary according to the research objective.

The large-scale survey system (LSSS, ver. 2.16.0) also has a global commercial presence and is marketed through a business unit of the Norwegian Research Center (NORCE) [20]. Its predecessor was the Bergen Echo Integrator (BEI) program [21], which already had automatic categorization. The echogram interpretation process of LSSS has been improved through multifrequency applications. The system is powerful because it rapidly interprets large volumes of data, taking 2 h to process a full day of the survey [20]. LSSS continues to be developed, so new features are constantly added. For example, it now supports the Sonar-net CDF4 format and can handle acoustic wideband data from the EK80 wideband echosounder. To get the most out of the program, the hardware requirement is computationally demanding: a 64-bit operating system and at least 32 GB of RAM.

Open-Source Software Options

Other initiatives have emerged from the scientific–academic community for acoustic data processing. Most of them have been built as open-source initiatives, a feature that allows the verification of the processing details and the creation of routines that can be shared inside and outside the workgroup, which help to foster collaboration. This software is more economical but less robust (Table 1).

Matecho (ver. 7) was developed in MATLAB by the Development Research Institute (IRD). It can be used by running MATLAB Compiler Runtime (MCR) or by using the Matecho standalone version, although it is always recommended to run it within MATLAB. This software works in conjunction with MOVIES 3D (ver. 2.2.7 and further) [22]. It has automated processing that allows important functions to be performed to estimate abundance, such as echo integration by layer, extraction by schools, and noise removal. Moreover, it includes algorithms for applying methodologies based on multifrequency filters [23]. These features provide similar advantages as those implemented by Echoview. An important feature is that it supports the HAC standard format, which was adopted by the International Council for the Exploration of the Sea (ICES) Fisheries Acoustics Science and Technology Working Group (WGFAST) in 1999 for the exchange of raw and edited fisheries acoustics data. Thus, it allows for the recovery of files that have been previously processed in that format or for the processing raw data, including those collected by the EK80 broadband echosounder. The hardware requirements are simple and easily accessible (64-bit Windows 7 or 10 and ≥4 GB RAM).

ESP3 (ver. 1.52.0) is an open-source software platform that can be downloaded at sourceforge.net. The algorithms used in the processes are visible, and the user can consult them in the software documentation. Similarly to Matecho, it is an executable program developed in MATLAB and can be used with MCR. ESP3 is free and powerful, but there are a few "bugs." These issues are solved by project administrators from the National Institute of Water and Atmospheric Research (NIWA). In the latest update (ver_1.52.0.), it is now possible to process multibeam echosounder data. ESP3 does not specify the hardware requirements, although it can run efficiently with 64-bit Windows 7 or 10 and ≥4 GB RAM [24].

StoX (ver. 4.0.0) is an open-source software package integrated with R. It was developed by the Institute of Marine Research (IMR) in response to the need for standardized software for research institutes working on marine studies in the North Sea, the Norwegian Sea, and the Barents Sea. The design features standardized sampling and repetitive protocols yet is flexible enough to allow workflows to be expanded, modified, deleted, and reorganized. Until 2019, StoX was used to estimate the abundance of stocks such as herring, sprat, blue whiting, cod, haddock, sandeel, and horse mackerel [25].

ECOPAMPA is a type of software created in Visual Studio 2010 to scrutinize synthetic echograms (implemented in C+ ver. 2.3). It was initially tailored for data collected by the National Institute of Fisheries Research and Development (INIDEP) using an EK500 echosounder. This software incorporates algorithms for the automatic detection of the seabed, surface, and schools of fish. It allows for the classification of schools of fish using artificial neural networks (ANNs). Of note, comparisons of experimental data between Echoview and ECOPAMPA have yielded good results. There is the potential to expand the application of ECOPAMPA to other regions, an endeavor that would require new data sets for training; however, this project is currently inactive [26].

Several projects in the Python (2024) programming language are at different levels of development; their advantage is that they do not have demanding hardware requirements. Perhaps the most ambitious and advanced is Echopype, a collaborative project that facilitates the integration of echosounder data into interdisciplinary oceanographic research [27]. Since 2019, its functions have increased due to multidisciplinary collaborations. Currently, Echopype supports data from the EK60, EK80, and Acoustic Zooplankton Fish Profiler (AZFP) echosounders, and it can convert raw data files into the Sonar-netCDF4 format [28]. Users with advanced programming skills can contribute via the GitHub repository, and those with limited experience with Python can find examples of how to use it. Other projects have been added, such as the PyEcholab library [29], developed in collaboration with different institutions in the United States to meet the needs of the National Centers for Environmental Information (NCEI). At this time, the code can read, plot, and export echosounder data. The EchoPy library (ver. 0.0.2) contains algorithms that allow for the basic processing of acoustic data, such as seabed correction and removal, the removal of corrupted pings, and multifrequency analysis (https://pypi.org/project/echopy accessed on 4 March 2024). A disadvantage of these libraries is their limited documentation. The long-term vision of these libraries is for the continuous contribution by other users to grow in a modular way, which would mean the expansion of the tool and accessibility for users beyond fisheries acoustic scientists.

Echogram (ver. 0.1.2), developed in the R language environment (version 4.2.3) [30], is a package that can manipulate multifrequency acoustic data in HAC and RAW (EK60) files and explore acoustic filters of echogram data (still under development [31]). Similarly to other software, Echogram can be run without demanding hardware requirements.

Due to the evolution of hardware, some processing software has become antiquated because of rapid advances in data technology: They have fallen into disuse or evolved into new versions. MOVIES+ (version is not reported), developed by IFREMER in the 1990s, was one of the first programs developed for fisheries acoustics data. It enabled the quantified analysis of echoes to manage fishery resources and ecological research. A particular feature was fish shoal extraction and characterization [32]. Over time, it has evolved into 3D MOVIES for the visualization of all echosounders in real time and in a three-dimensional (3D) environment. This software can incorporate data obtained from other instruments and is currently used by the French oceanographic fleet. Likewise, new software has been developed, including ESP3, ECHO (Division of Marine Research, Australia), SCHOOLS (Limnological Institute, University of Konstanz, Baden-Württemberg, Germany), SCHOOL (Institute of Marine Biology, Crete, Greece), CH1 and CH2 (Department of Fisheries and Oceans, British Columbia, Canada), and SCHOOLBASE (Marine Laboratory Aberdeen, Scotland) [33]. However, we have not considered them in this review because they are not used or there is limited available information.

The use of processing software in Mexico has varied over time. In the 2000s, IMIPAS began collaborating with the Research Institute for Development (IRD) and IFREMER to use MOVIES+; it was used from 2008 to 2018 to process the national acoustic surveys. However, this software cannot be installed on computers running Windows 8 and later. Consequently, since 2015, IMIPAS has invested in two Echoview licenses to process the massive amount of data collected from the water column. However, this investment was insufficient for the research vessels (RVs) that carry out acoustic monitoring in coastal areas in Mexico. To solve this problem, the first approach was to use free software; Echogram was the first choice. However, it has limitations in processing massive amounts of data. The second approach was to use Matecho, whose main advantage is its ability to analyze data from entire surveys. This program was used for 1 year. For the past 3 years, ESP3 has been used.

In Mexico, the use of different criteria and subjective user decisions make it nearly impossible to obtain the same output from two or more different processing software programs. This issue is exacerbated for long time series and the effects of tests or new methods. A single processing program is often tailored to the needs of the institution and used to establish a workflow and to improve the software. Hence, institutions with well-established acoustics programs maintain a workflow for analyzing their data, and their monitoring has improved continuously. Historical transitions among processing software programs used in Mexico have not favored the reproducibility and consistency required for acoustics research. For example, MOVIES+ cannot provide robust and accurate estimates of background noise, while ESP3 contains algorithms for these estimates. Similarly, Matecho was tailored for the acoustics data collected by the French oceanographic fleet, and this implies difficulties when implementing it for different data sources. Moreover, slight changes in results can have significant management implications (e.g., individual quota and the harvest rate).

Free software remains in a stage of development and is maintained based on the interests of the public institutions that finance it, so it is unknown how long it will continue to be developed. Given the uncertainty around the availability of processing software, it is difficult to train specialized technicians effectively. The final choice of the processing software could be based on several factors: the possibility of a decrease in the budget that covers commercial license renewals, the possibility that free software will become outdated in the short term, and the possibility that software will be unable to support new, updated acoustic instruments.

Available processing software may only be endorsed by their institution or program manager, and many lack standardization or even documentation. Hence, the main issue is to establish whether there are differences among the available software. A desirable focus could be on the quantitative comparison of the following features: (i) the capacity to analyze a massive amount of data provided from multiple frequency channels; (ii) the suitability and accessibility of the programming language; (iii) suitable documentation indicating the codes used and a user manual; (iv) evaluation and comparison of the algorithm's performance; (v) the use of the metadata to estimate the volume backscattering coefficient (e.g., calibration values, pulse duration, and transmitting power) and the impact of metadata on the estimation of the area backscattering coefficient and nautical area scattering coefficient; and (vi) comparison between outputs of mean density and abundance because both quantities are relevant for fishery management (i.e., management quantities that are used to advise stakeholders).

When the power of each processing software is known and the methods are sufficiently general, the software can transcend the quantitative estimation of fish abundance and be used for a wide range of topics. We highlight the importance of using processing tools that are widely accessible and used in the scientific community. Ecologists, freshwater researchers, and others should be trained to use these tools. The accessibility of the processing software is not a problem for countries with a tradition in acoustics and that have achieved consolidated working groups. However, technology transfer has been a problem in Mexico. Although the country is on an upward curve to consolidate the application of acoustic techniques, the availability of processing software has become a "bottleneck". Breaking this bottleneck is the most immediate challenge for Mexico.

Table 1. The most cited software available for processing acoustic data.

Software	Open Source	Project Administrator	Country of Origin	Programming Language	Availability	References, Contacts, and Repositories
Echoview	No	Echoview	Australia	C++	Request sales information from info@echoview.com	info@echoview.com
Echopype	Yes	The project is directed by Wu-Jung Lee and Emilio Mayorga, but there are other collaborators	United States	Python	https://pypi.org/project/echopype/, accessed on 16 July 2024.	https://github.com/leewujung https://github.com/emiliom leewj@uw.edu, accessed on 16 July 2024
Large-Scale Survey System (LSSS)		MAREC—Institute of Marine Research (IMR)	Norway	Java	Request sales information from info@marec.no	[34]
ESP3	Yes	National Institute of Water and Atmospheric (NIWA)	New Zealand	Executable with a MATLAB license or with MATLAB Compiler Runtime (free)	https://sourceforge.net/, accessed on 4 September 2024 * You can subscribe to project updates to receive notification of new versions	Yoann.ladroit@niwa.co.nz Pablo.Escobar@niwa.co.nz
StoX	Yes	Institute of Marine Research (IMR)	Norway	Java	https://github.com/StoXProject/StoX/releases/tag/v4.0.0, accessed on 20 July 2024	espen.johnsen@hi.no arnejh@hi.no
PyEcholab	Yes	Institute for Research in Environmental Science, University of Colorado, Boulder, and collaborators	United States	Python	https://github.com/CI-CMG/pyEchola/, accessed on 20 July 2024	wcd.info@noaa.gov

Table 1. *Cont.*

Software	Open Source	Project Administrator	Country of Origin	Programming Language	Availability	References, Contacts, and Repositories
EchoPy	Yes	Institut de Recherche pour le Développement (IRD)	France	Python	https://pypi.org/project/echopy/, accessed on 12 August 2024	echopy@protonmail.com
ECOPAMPA	Yes	The Physics and Engineering Research Center of the Center of the Province of Buenos Aires (CIFICEN)	Argentina	Visual Studio 2010		svillar@fio.unicen.edu.ar
Matecho	Yes	Institut de Recherche pour le Développement (IRD)	France	Executable with a MATLAB license or with MATLAB Compiler Runtime (free)	https://git.outilsis.ird.fr/activeacoustics/matecho, accessed on 19 August 2024	[22]
MOVIES + Movies-B	No		France			[32,35]
Echogram	Yes	Centro Interdisciplinario de Ciencias Marinas	Mexico	R	https://github.com/hvillalo/echogram, accessed on 26 August 2024	hvillalo@inp.mx
Bergen Echo Integrator (BEI)	No	Institute of Marine Research (IMR)	Norway	C++		[21,36]
SONAR 4,5,6	No					
MOVIES 3D	No	L'Institut Français de Recherche pour l'Exploitation de la Mer	France		http://flotte.ifremer.fr/Presentation-de-la-flotte/Logiciels-embarques/HERMES-MOVIES3D/Telechargem, accessed on 2 September 2024	

4. Sampling Design

All known information on a population should be used to create an efficient sampling design that provides the best field data to estimate biomass. For this purpose, the placement of transects must maintain a balance between the total area to be covered, the available time, and the assigned budget. For Mexico, the objective is to design a survey that generates a long and consistent time series, because once the design is chosen, it can be used systematically to produce estimates that are comparable to each other.

Classical sampling theory states that a random sampling procedure provides unbiased estimators of the mean and variance. If the sampling design used is systematic, the samples are considered not to have been taken independently of each other, and statistically, there is no valid estimator of variance from a systematic sample unless the population is randomly distributed [37]. The schools of fish vary in two dimensions, height and width, and the closest observations within a sample could exhibit a high "correlation" because this factor depends on the distance between the observations [38]. If sampling is performed at a scale larger than that of the correlation length, classical methods can be used to estimate the variance. But if this is not the case, then a model that explores the possibility of autocorrelation of the samples is needed.

In acoustics, three commonly used sampling designs are systematic zig-zag, systematic parallel, and stratified random parallel. The stratified parallel transects provide maximum separation between transects, resulting in the least correlation between them and greater coverage. However, it may be more difficult to estimate the variance of the estimate. According to Kimura and Lemberg [39], zig-zag transects are better than stratified parallel transects at low transect densities; on the contrary, stratified parallel transects are better than zig-zag transects at higher transect densities. Furthermore, Kimura et al. [40] explained that change regions are oversampled with the zig-zag design, while regions between changes are undersampled.

Mexico uses these three sampling designs in different regions, covering the area of distribution of the small pelagic fish populations in northwestern Mexico, including the sampling of shallow bays, which have historically maintained a substantial portion of populations of these species. The Gulf of California and the west coast of the Baja California Peninsula use a stratified parallel sampling design, while a zig-zag design is used for the mouth of the Gulf of California (Figure 2), considering a sampling design until the 200 m isobath that has been surveyed in the recent 15, 12, and 10 years, respectively, using the same echosounder, ship, and fishing protocol [41–43].

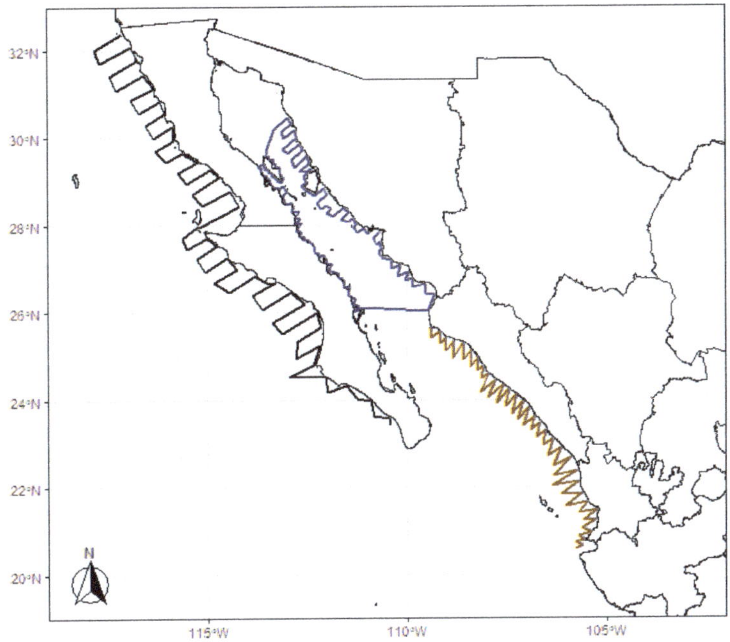

Figure 2. The sample design for acoustic survey research indicating the transects used in the different regions of the Mexican northwest Pacific Ocean. The three main regions are spatially located as follows: (i) the west coast of the Baja California Peninsula (transects in black); (ii) the northern region of the states of Nayarit and off the shore of Sinaloa (transects in gold); and (iii) the Gulf of California (transects in blue).

Fréon et al. [44] mentioned that schools of fish display migratory behavior at night-time toward shallower waters, where they lose their ability to behave as schools and disperse and extend into aggregations and layers in search of food. In northwestern Mexico, pelagic fishing activities occur during the night and the so-called el oscuro, a period of 22–26 days centered on the new moon of each month [45]. In the Gulf of California, schools are maintained while they migrate to shallower waters, and they are distributed closer to the surface (<40 m) at night [43], making them more accessible for commercial fishing and sampling.

This school behavior requires a different acoustic sampling procedure implemented by agencies such as NOAA (California, United States) for Pacific sardines; IFREMER (Brest, France) for anchovies; MEDIAS (Mediterranean International Acoustic Survey) for European anchovy and European sardine [46]; the Japan Fisheries Research and Education

Agency (FRA) for anchovies, herring, and other pelagic species; and the Institute of Marine Research (IMR) for herring [47]. For these species, the estimates of density and abundance are mainly supported by acoustic data collected during the day. Therefore, the night-time acoustic datasets are neglected due to diel vertical migration of the species, avoiding negative bias in the outcomes. Consequently, these night-time data provide little information on the stock status [48–51]. During the night, the acoustic sampling performed by these agencies can identify species by superficially using mid-water trawls. This procedure implies that the ship returns to previous positive areas where the target species were identified with acoustic equipment. In Mexico, when a school is identified during a survey, the ship keeps going until it has gone 1 nautical mile, which is an elementary sampling distance unit. Then, the transect is interrupted to carry out the fishing set in the opposite direction. At that time, the captain and the crew execute the fishing maneuver to position the net at the depth of the previously observed shoal. Because the ship lacks net sensors, the angle at which the net is deployed is monitored by the researchers. The fishing set begins with a mid-water trawl for 30 min at a constant speed of 3 knots. Subsequently, the ship goes back to the point of interruption to provide continuity to the survey. The criteria for carrying out the fishing sets are described by Doray et al. [50].

The presence of new platforms for survey research in Mexico has allowed the implementation of alternative methods for the acoustic data collection for small pelagic species. For example, since 2021, the RV Dr. Jorge Carranza Frazer has used a methodology closer to the acoustic data collection proposed by NOAA. During 2023, there were changes to the sampling design at the mouth of the Gulf of California: the surveyed transect was modified from zig-zag to stratified parallel, the EK60 echosounder (38 and 120 kHz) was replaced with a broadband frequency echosounder (EK80) with a single frequency (38 kHz), and the estimates of density and abundance were mainly supported by daytime data collection and nighttime mid-water trawls. Stock assessment and fishery management approaches often use time series that are collected during research surveys with standardized protocols (the number of sampling stations, similar fishing gear over time, and defined periods). These changes represent breaks in the time series because direct comparisons between estimates might not be easy.

5. Frontiers for the Application of Acoustics

5.1. The Contribution of Acoustic Data to Ecosystem Studies

The scientific community agrees that to maintain sustainable fisheries, it is not sufficient to assess the target species alone; the role of exploited species must also be assessed within the ecosystem in which they live [52]. The implementation of this management approach requires multidisciplinary efforts that include multiple species and synoptic and periodic monitoring to promote more realistic fishery management.

The first effort made to understand the structure and function of marine ecosystems through the trophic relationships between components of the studied communities was the direct application of "ecosystem models" developed in the early 1980s [53], which were later adapted to routines such as Ecopath, Ecosim, Ecospace [54], the integrated software package EwE [55], and Ecotroph [56]. These tools have been useful in addressing questions regarding the ecosystem impacts of fishing. However, these models have a high degree of complexity that increases as the number of functional groups increases. In addition, various types of information, including species abundance, diet composition, catches, consumption rates, ecosystem properties, and trends of species biomass, are required for their application to plausible scenarios [57]. In commercially exploited species, discard and catch information from arrival notices can provide that information or even help to corroborate the results to reinforce the time series, but species that are not commercially exploited are subject to unverified assumptions. For example, there are no catch data available for the "mid-trophic level" functional group, which includes cephalopods, gelatinous organisms, adult euphausiids, and mesopelagic fish [58]. Nevertheless, they represent an important group, particularly in ecosystems with wasp-waist-type dynamics [59]. If ecological management units are spatially large, they will never be assessed properly due to the limitations associated with human and economic resources [60].

Comparatively, acoustic data provide quantitative and qualitative metrics over a wide range of scales and with high spatiotemporal resolution [52]. They allow for observations ranging from hours to long time series, and they can gather information on marine organisms of all sizes (from fish eggs and larvae to zooplankton and larger species), identifying and estimating them by size. Furthermore, they provide data on scales ranging from centimeters to ocean basins [61].

According to Bertrand et al. [62], ecosystem models do not consider the variability and the interactions between the physical environment and species communities. According to Lazzari and Tupper [63], when the complexity of the habitat increases, properties such as its species richness and abundance also increase. Other properties associated with the seabed also play an important role in the habitat description [64]. In this sense, acoustic methods can provide data such as substratum characteristics. By associating the backscatter intensity with the angle of incidence, it is possible to detect differences in seabed properties [65,66], and the frequency response can indicate the type of seafloor [67]. For example, Siwabessy et al. [68] performed a backscatter signal analysis to determine the backscatter characteristics of seabeds associated with seagrass beds. Cutter and Demer [67] used multifrequency split-beam echosounders and proposed the multifrequency biplanar interferometric imaging technique to predict the potential habitat for demersal fish (*Sebastes* spp.); this finding helped to optimize the surveyed area. Gastauer et al. [69] identified different seabed habitats to investigate the ecological meaning of clusters of fish and the distribution of fish density hotspots.

There are differences in the level of implementation of ecosystem-based fishery management (EBFM) between developed and developing countries [70]. In Mexico, most management strategies are aimed at a single objective. Ecopath was constructed for the northern Gulf of California, including for small pelagic species [71], and for a benthic ecosystem exploited by shrimp trawlers in the Gulf of California [72] and the Gulf of Ulloa [60], and it has been applied to jumbo squid (*Dosidicus gigas* [73]) and loggerhead marine turtles (*Caretta caretta* [74]). The available Mexican fisheries information is quite variable; for a few resources, there are more than 50 years of well-documented information, while for others, in the best of cases, there are basic levels of information. These fisheries usually make a minimal economic contribution to the region; consequently, they have a limited research budget. Applying ecological models such as Ecopath to these resources is not a viable option, because the detailed biological fishery information may be limited or even unavailable.

Currently, acoustic data collected for stock assessment are not utilized effectively in understanding the interactions among species. The contribution of information from plankton and micronekton on ecosystem functioning or the study of ecological relationships, such as prey–predator and intra- or inter-specific competition, is not considered. During data processing, these are assumed to be noise and therefore eliminated. In Mexican research, ecosystem studies supported by acoustic tools have been developed to a limited extent. Few studies had been published prior to 2020 [43,75,76]. More recently, collaborations between national and foreign ecologists have flourished. One example is the daily vertical migration of zooplankton and its seasonal cycle in the Gulf of Mexico [77]. In the Gulf of California, there have been some recent ecological studies supported by acoustic data. Portner et al. [78] analyzed the responses of the mid-trophic communities to oceanographic and climatic variability, while Sarmiento-Lezcano et al. [79] estimated the biomass of migratory mesopelagic species, decapods, and euphausiids and their role in the respiratory flux. However, the technique is not widely used in Mexico. Other Mexican researchers in marine sciences only use acoustic data to evaluate resources.

There are an increasing number of publications that have used an ecological approach with acoustic methodologies throughout the world, indicating the appropriation of acoustic techniques in diverse research fields. In Mexico, to avoid failures in the potential application of acoustic methodologies, the advantages of acoustic techniques need to be disseminated and socialized so they that are more understandable to researchers before these tools can be incorporated into research programs. In addition, these are perceived as expensive tools that are used only for industrial fisheries, while most of the fishing in Mexico is carried out with coastal or small-scale fleets. Collaboration between researchers from different disciplines could facilitate the combination of acoustics as a tool and ecology as the main objective to enhance the scenarios that are moving toward EBFM by integrating multidisciplinary oceanographic research programs, ecology, and fishing. Acoustic data from a range of platforms could provide auxiliary and complementary information to develop better ecosystem models.

5.2. The Relevance of Acoustic Data for Marine Protected Areas and Endangered Species

Marine Protected Areas (MPAs) represent a management tool created as an alternative approach to reduce and prevent the intensity of human pressure on marine life and habitats [80]. Mexico has decreed 66 MPAs covering an area of approximately 13.12 million ha. The details of the geographic distribution of these MPAs are available in a previous study [81]. Since their implementation, some MPAs have been more successful than others. For example, in the Gulf of California, ~23,300 km^2 have been decreed as MPAs, but only the national park known as Cabo Pulmo has met conservation and sustainability goals [82]; it has experienced a great recovery in the biomass of fish and top predators [83].

There are guidelines and recommendations for MPAs, which highlight the need for long-term national programs to monitor variations in key parameters such as the age structure of marine organisms, variations in the number of prey, and the rate of habitat loss [84]. Most of the literature on MPAs involves traps or baited traps, fishing, hook-and-line fishing surveys, optical methods, and underwater visual censuses with divers [85–87]. While these methodologies are useful for bottom-dwelling species, there is a lack of knowledge on the response to the management measures for pelagic and benthic species [88].

Acoustic methods could be a particularly valuable tool for these regions. The procedures are non-invasive, can provide long-term observations and sample species assemblages throughout the water column over a wide spatial range in a record time, and may even be more economical than the methods previously described [85]. Furthermore, these methods can be combined with other tools—for example, optical methods to identify species mixtures and their size structures in areas where fishing is restricted [89,90]—even when the rocky high-relief substrates are inaccessible to standard survey trawls [91]. Combining acoustics with systematic conservation planning tools can take monitoring a step further, allowing for the adaptive management of protected areas to assess reserve effectiveness based on the spatiotemporal patterns of ecosystems [92].

Considering the number of MPAs in Mexico and the budget to evaluate them, the selected tools must be able to monitor the recovery of fish stocks effectively. Mexican scientists have published innovative work on this topic. Mayorga-Martínez et al. [93] showed that the planning and management of marine areas in the country could benefit from the application of active acoustics. They evaluated the fine topographic complexity of a group of coastal coral reefs in Veracruz at mesophotic depths (>30 m), which are rarely included in marine reserve designs and management, despite their ecological importance and connectivity to shallow reefs. With this information, they proposed expanding and reshaping the core zone to include the high structural complexity exhibited by the entire reef complex. Another example of sampling to monitor change inside MPAs was carried out in the El Bajo Espíritu Santo seamount. Villalobos et al. [90] described the bathymetry, oceanographic habitat, distributions of zooplankton and fish, determination of the dominant fish species, estimation of their biomasses, and determination of the target strength (TS) of the Pacific creolefish (*Paranthias colonus*) and the finescale triggerfish (*Balistes polylepis*). The researchers obtained these results by collecting and integrating acoustic, optical, and oceanographic data.

Mexican scientists have also used innovative approaches to apply acoustic techniques for endangered species. The Biosphere Reserve of the Upper Gulf of California was established as an MPA with the purpose of protecting vulnerable species. Since its creation, various social problems have led to an active debate about fishing in the area [94]. The totoaba (*Totoaba macdonaldi*) is an endemic species of the Gulf of California. Since 1976, commercial fishing of this species has been banned due to the alarming decrease in its stock. Nevertheless, illegal fishing and a lack of scientific monitoring have led to controversies regarding stock conditions and conflicts of interest in the management of the current population [95]. In this region, the most recent update on the current totoaba biomass was supported by fishery-independent data (an acoustic survey).

As MPA management becomes more effective, the effects will be reflected in the enhanced provision of a variety of ecosystem services for the communities. In the Mexican context, these kinds of studies require institutions to apply innovative methodologies that produce the best results and that provide stakeholders with the best advice for their decision making.

5.3. *Acoustic Assessment of Fishery Resources in Freshwater and Shallow Water*

Inland waters are defined as lakes, rivers, streams, canals, reservoirs, and other land-locked waterbodies. They have an important role in providing animal protein for humans, in maintaining riverine fish abundance, and have an ecological role as a reservoir of biodiversity [96]. Fishery hydroacoustic research in freshwater aquatic ecosystems has mainly been developed in North America (Canada and the United States) and northern and central Europe. Its development has been favored by significant reductions in the size of echosounders and other acoustic instruments as well as cost reductions. Other important work has been undertaken in Australia, China, and the East African Rift Valley. Conversely, there have been fewer studies from South America, Russia, and New Zealand. This growth is parallel to advances in marine acoustics, which are quickly being adopted by developed countries [97].

Mexico has about 320 hydrological basins. Among these, there are 37 main basins, of which 12 discharge into the Gulf of Mexico and the Caribbean Sea, 19 that discharge into the Pacific Ocean and the Gulf of California, and 6 that are endorheic. It is estimated that there are about 70 lakes in Mexico. There are also 840 reservoirs classified as large dams. In 169 of them, there are productive activities such as commercial fishing, sport fishing, and fish farming. The scientific research that is carried out on species of commercial importance in freshwater ecosystems is of a fishery biology type (e.g., yield per recruit and catch per unit effort (CPUE)). In contrast, there are no periodic evaluations of biomass for a large portion of these reservoirs. According to the National Fisheries Chart [19], the most important species for commercial fishing are the groups of mojarras (i.e., cichlids such as tilapia), carp, channel catfish, and charales; in sport fishing, the species of interest is the sea bass.

In Mexico, Linares et al. [98] applied acoustic techniques to evaluate red grouper aggregations, recognizing the potential of these applications in the evaluation of fish that inhabit areas of steep slopes or use them as spawning grounds. This work is pioneering in the modern application of these techniques in southern Mexico. It highlights relevant points to improve the results in the application of acoustics in shallow waters, which can be applied to other areas of the Mexican Caribbean.

Rowell et al. [99] estimated the abundance, biomass, length structure, and spatial distribution in the Gulf Corvina in the Colorado River delta, Mexico, through several acoustic surveys combined with hydrophones (passive acoustics). This procedure provided excellent fishery-independent data. Besides that study, little research has been carried out in shallow waters, despite the fact that a large part of the territory has these characteristics.

Similarly to the situation with some MPAs in Mexico, the budget available to evaluate dams and lakes must be maximized as much as possible. In large dams, the use of nets represents a passive method that increases the time required to obtain the dataset.

Moreover, there is a high labor cost, and the specific fishing gear chosen could bias the biological data, mainly the size structure of the analyzed fish population, leading to a lack of continuity [100]. On the other hand, the range of direct visual underwater observations is limited to a few meters because in many Mexican waterbodies, the optical properties are poor and transparency is low; thus, the use of other optical sensors, including satellite images, is very limited. Acoustic methods offer new possibilities for evaluation. Echosounders can be easily installed on small boats, and institutions with experience and instruments can provide training so that a small team of local researchers can operate hydroacoustic technologies.

5.4. Target Strength

TS is a scaling factor to convert the intensity of the energy values that are backscattered by the object into abundance. Without knowledge of an organism's TS, it would not be feasible to translate acoustic densities into biologically meaningful data. TS is not a single value; rather, it is the average value of a distribution of individual observations of the backscattering cross-sections (σ_{bs}, with the units m^2) that are often expressed in terms of body length [101]. There are extensive reviews on TS measurement methods [102–104]. These experiments can be separated into in situ—using measurements made on fish with a natural behavior and free swimming—and ex situ—using measurements made on fish confined to a cage, immobilized, or unconscious. The checks are influenced by a large number of factors including the method used, the angle of inclination of the organisms, and up to 90% of the state of the swim bladder (if any) [105]. In addition, the estimated values are not interchangeable between the frequencies used.

TS is considered to be the most important source of error in acoustic estimates of absolute abundance [101]. Underestimating the TS would overestimate the abundance, while overestimating the TS would underestimate the abundance. Hence, due to their economic importance, considerable efforts are made to determine the TS of species that support fisheries. Of note, research on this topic is also growing for organisms of ecological importance in disturbed ecosystems. TS has been used widely to estimate species distributed in the North Sea, the Baltic Sea, the North Atlantic, the Southeast Atlantic, the North Pacific, and even the South China Sea (see Table 4 in Liu et al. [106] and Table 6.4 in Simmons and MacLennan [101]), among others. In tropical species, this type of study is limited. When the TS for any species is unknown, published values of similar species or the same species distributed in another region are usually used. However, this approach introduces inaccuracy, which reduces the potential of the data obtained from survey research by affecting estimates of the relative abundance and biomass.

For Mexico, there have been very few specific studies estimating TS. The first efforts focused on estimating the TS of species distributed in Mexican waters were for Pacific thread herring (*Opisthonema* spp.), totoaba, and giant squid (*D. gigas*). Although Mexican researchers perceive that this issue is important, these TS models are insufficient for a country that has large commercial and potential fisheries. Ideally, acoustics research must include the development of TS models, especially in those cases where there are no TS models in the literature to provide support for acoustic survey data. In the next decade, Mexico must focus on this critical point to implement and consolidate the technique in the country.

6. The Relevance of Time Series

6.1. Proportionality between Indices of Relative Abundance and Abundance

A stock assessment can usually be supported by two kinds of datasets. The first comprises fishery-dependent data (i.e., extensive data collected by commercial fleets). The second comprises fishery-independent data, which are commonly obtained from survey research cruises. There are fundamental requirements for these data to become valuable: a consistent year-to-year sampling design that covers the distribution of the stock, a carefully standardized fishing operation [107], and fishing nets with selectivity that can be assumed to be constant [108]. For any commercial fishery, the catch rate or CPUE is a traditional index of relative abundance. Thus, the CPUE may be used as a proxy for the stock abundance, assuming that both the CPUE and stock abundance are proportional [109]. However, it is usually difficult to make this assumption. For small pelagic species, the CPUE could remain relatively stable while the stock abundance decreases. This phenomenon, known as hyperstability, indicates a disproportionate relationship between the CPUE and stock abundance [109,110]. Another phenomenon for commercial fisheries is the density-dependent catchability, which occurs when the catchability of small pelagic fish schools increases as their stock abundance decreases [110–112]. From this perspective, hyperstability and density-dependent catchability exhibit a mirror effect for measuring the fall in stock abundance and both mask a decline in the index of relative abundance time series (e.g., CPUE) (Figure 3).

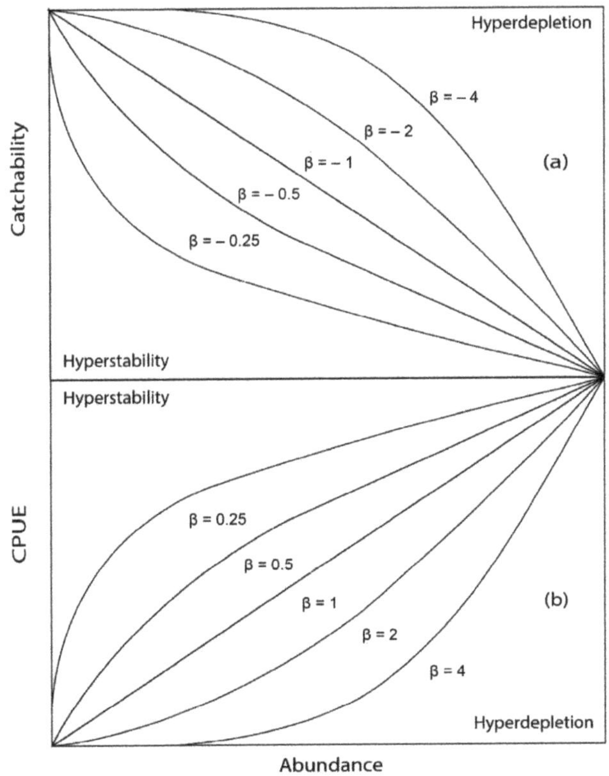

Figure 3. The graphs show (**a**) the relationship between abundance and catchability, indicating patterns of density-dependent catchability, and (**b**) the relationship between abundance and catch per unit effort (CPUE), exhibiting different patterns of hyperstability [109,110,113].

6.2. The Relevance of Density Time Series and Their Use as an Index of Relative Abundance

Since the 1980s, fishery models fitted with multiple time series using fishery-independent data have been implemented successfully [114,115]. Indeed, even a "weight factor" has been added to these data to measure the performance of fishery-independent data, thus enhancing the estimation of parameters in the models. The main reason for using multiple time series is related to the quantity of information contained in each dataset. For example, a time series of an index of relative abundance that exhibits a negative trend indicates that the population abundance is decreasing. In contrast, a different index of relative abundance could indicate a positive trend for the same time series; consequently, this index would suggest that the population abundance is increasing. It is a difficult situation when these contradictory datasets are recorded and included in the fishery models [113]. However, there is also an advantage: multiple time series could contain complementary information that helps to understand the changes in abundance. In Mexico, although independent estimates of fisheries began with an assessment program of eggs and larvae for pelagic fish stocks, they were eventually discontinued due to budget limitations. In 2008, the acoustic methodology was implemented and replaced the abundance estimate provided by the collection of eggs and larvae of small pelagic species. In other countries, abundance estimates using eggs and larvae have been combined with acoustic methods, and the generated time series are used in population assessment models. During the generated time series, there are sometimes changes in the fishing gear or ships used and even increased sampling with other tools. Because the new gear has its time series, the old and new time series can be entered as separate indices of relative abundance or statistically standardized [116].

MacLennan et al. [117] highlighted that density (ρ) is proportional to the area backscattering coefficient (s_a). This assertion suggests that s_a or the nautical area scattering coefficient (s_A) could be used as an index of relative abundance and, consequently, to assume that it is proportional to abundance (*N*). Assuming that a time series of ρ values has been estimated, it could be possible to assume a priori the changes in the abundance of biological targets.

According to [109], the index of relative abundance (*I*) in trawl surveys can be expressed as $I = qN$; consequently, *N* can be estimated as $N = \frac{I}{q}$, where q represents the catchability [116,118]. The main assumption is that *I* is proportional to *N*. For this reason, estimating the best *I* time series is difficult. The best option is to utilize several *I* values. If the proportionality for any *I* is weak, then the equation could be re-expressed as $I = qN^{\gamma}e^{\varepsilon}$, where γ is a power parameter indicating proportionality when its value is 1 or close to 1, and ε is a random error assumed as $\varepsilon \approx N(0, \sigma_{\varepsilon})$ [109]. MacLennan et al. [117] explained that acoustic studies provide two important quantities: the density and the abundance of biological targets. Specifically, ρ is expressed as the number per unit surface area of the layer. An estimation of ρ could represent an index of relative abundance for an area at time t. Perhaps the most important mathematical expressions using two surface scales are $\rho = 10^6 \frac{s_a}{\sigma_{bs}}$ for km^2 and $\rho = \frac{s_A}{(4\pi\sigma_{bs})}$ for nmi^2, where σ_{bs} represents the backscattering cross-section. The assumptions about the field data, including the statistical distribution (e.g., normal, lognormal, and delta), the presence or high frequency of negative sampling stations, and a spatially structured population, would affect ρ values—they could be biased and would provide little information about the biomass. To avoid a failure to estimate ρ, several assumptions may be made depending on the methodological approach: the application of generalized additive models (GAMs) with spatial variability, assuming that the samples collected are statistically independent of each other; the application of geostatistical techniques that can represent the spatial variation of density, providing solutions to the spatial correlation in a structured population without any assumption regarding the independence of the sample; and a lognormal/delta distribution, assuming the effect of negative sampling stations on estimates of ρ (e.g., Pennington estimator), also assuming statistical independence in the samples.

The random functions for interpreting the spatial distribution of a population within a specific area allow for the development of models with explicit spatial variability, such as generalized linear models (GLMs) or GAMs [119]. One of the main disadvantages of these methods for abundance estimation is that they do not provide simple estimates of error variance or survey precision.

There are a large number of geostatistical tools that allow one to determine the sample variance in a systematic sampling design. The most relevant aspect of a geostatistical approach, and perhaps why it is currently one of the most widely used approaches, is that it does not require the independence of the sample to be assumed, and it offers an explanation for spatial variability [120].

The delta-lognormal approach proposed by Pennington [121] includes highly skewed data (known as positive sampling stations) and negative sampling stations denoting a value of zero (apparent absence of the target species). Both datasets are used to estimate the mean value of the survey. For Syrjala [122], this method is not resistant to relatively small and undetectable deviations from the assumptions of the model. Thus, an excess of small values generates poor behavior in the delta-lognormal distribution, an increase in the bias, and a loss of information in the estimation of the mean value [40,123].

In Mexico, these topics have not been widely analyzed by fishery biologists, but they have been applied in other areas such as earth sciences and engineering. Thus, the integration of interdisciplinary groups regarding the use and application of these statistical procedures would be desirable. There is the possibility that the acoustic samples have spatial autocorrelation; the degree of this autocorrelation depends on the size of the schools (length and width) and the distance between transects. Even if these interactions are ignored, they would impact the estimated variance.

6.3. Hydroacoustic Data and Integrated Stock Assessment Models

Traditionally, the fishery management approach used by international fisheries agencies (e.g., NOAA, IFREMER, the General Fisheries Commission for the Mediterranean (GFCM), and the Scientific, Technical and Economic Committee for Fisheries (EC STECF)) has been based on abundance estimates comparing biological hypotheses and datasets. NOAA has developed two methods to measure the uncertainty and discrepancies in outcomes such as abundance as well as management quantities: the explicitly vulnerable biomass and the annual harvest rate. The most widely analyzed stocks have been small pelagic species, mainly the Pacific sardine and the northern anchovy.

The first method is based on ichthyoplankton surveys and extensive field datasets collected using a stratified sampling design [124]. The datasets were analyzed using the DEPM, which is based on the ratio between daily egg production and the specific daily fecundity of the species [125,126]. This approach provides estimates of the biomass, spawning biomass, and related biological parameters, as well as indices of relative abundance such as the egg production index (eggs 0.05 m^2 day^{-1}), DEPM data (10^3 t), and historical egg production data (eggs 0.05 m^2 day^{-1}) [127]. These indices have been used frequently for stock assessment purposes. Moreover, additional indices of relative abundance have been computed, such as the reproduction area (nm^2) and the proportion of positive stations, which refer to the marine stations where the eggs or larvae of target species were collected [128,129].

The second methodological approach is the application of stock assessment models, mainly those called integrated catch-at-age models. These statistical models can represent the most important features of population dynamics (i.e., growth, recruitment, and mortality, both natural and fishing) and incorporate all the information known about the analyzed species. When these models are tuned using fishery-independent data, their realism increases. In the last two decades, the implemented methodologies have been diverse (Table 2).

Acoustic data can also provide an estimate of the biomass and indices of relative abundance and can serve as an alternative to the DEPM. Moreover, ρ and S_A have suitable statistical properties for use in the integrated catch-at-age models [109]. This manner of assessing marine resources has been a standard procedure. Comparing outcomes that search for the best scientific evidence can increase the level of scientific discussion, enhance the conclusions, and help to maintain the sustainability of the target species.

In Mexico, the DEPM has been implemented successfully [18,130], but it is not used systematically. Hence, there is a lack of consistency in the temporal sequence of the obtained outcomes, and the time series of the biomass and the index of relative abundance are only used in the stock assessment models. Since 2008, acoustic data have been collected systematically in the Gulf of California, where the target species is mainly the Pacific sardine. Estimates of the biomass are available for 15 years (2008–2023). The index of relative abundance used to fine-tune the integrated catch-at-age model of the Pacific sardine in the Gulf of California is biomass-computed from the acoustic data [131]. The rationale and support for this procedure are based on the fact that the time series of acoustic estimates do not include a relative and time-dependent proportion of the real abundance of the target species. This value is known as detectability [132]; it is estimated within the insonified volume, included in the echo-integration, and can reduce bias in the final estimate. Therefore, the biomass calculated from acoustic data can be understood as an index of relative abundance rather than an absolute value of biomass [132]. Detectability has been estimated for benthic resources, but its utility for small pelagic species remains unclear.

In Mexico, stock assessment for small pelagic species has focused on maintaining two sources of information for fishery management purposes: one based on acoustic data and the other applying fishery models fine-tuned through acoustic outcomes [42,131]. Future challenges include choosing the best indices of relative abundance for the stock assessment models, the possibility of incorporating the density estimated from acoustic data, and the use of different density values considering the spatial distribution of the target species in the Mexican Pacific Ocean. Finally, the objective is to use more informative fishery-independent data—acoustic data may be a good choice. Changing the input data is a common procedure when integrated catch-at-age models are solved. In this way, different assumptions can be analyzed, and the performance of the fishery models can be assessed. The contribution from scientific fishery management will have an impact on the advice to stakeholders and decision makers, possibly affecting the governance of the target species.

Table 2. Age-structured models are commonly used to input data such as acoustic data (including indices of relative abundance). The models included are the most used and referenced in the worldwide literature.

Package Name	Acronym	Does the Package Include a Population Dynamic Structure?	Is Uncertainty Assessed?	Does the Model Require Indices of Relative Abundance?	Is Documentation Available in the Form of a Peer-Reviewed Publication?	Source
Age-structured Assessment Procedure	ASAP	Yes	Yes	Yes	Yes	[133]
C++ Algorithmic Stock Assessment Laboratory	CASAL	Yes	Yes	Yes	Yes	[134]
Stock Synthesis Model	SSM	Yes	Yes	Yes	Yes	[114,135]
Assessment Method for Alaska	AMAK	Yes	Yes	Yes	Yes	[136]
Simple Stock Synthesis	SSS	Yes	Yes	Yes	Yes	[137]
Extended Simple Stock Synthesis	XSSS	Yes	Yes	Yes	Yes	[138,139]
Woods Hole Assessment Model	WHAM	Yes	Yes	Yes	Yes	[140]
A Length-based, Age-structured Model	Multifan-CL	Yes	Yes	Yes	Yes	[141]

7. Mexican Budget for Research

7.1. Acoustic Equipment and Instrumentation: Mexican Research Vessels

In Mexico, there are currently three platforms that aim to carry out marine studies and acoustics research. Two of them are the property of the IMIPAS, which coordinates and provides scientific research and advice in fisheries and aquaculture. The platforms are the RV BIP XI, a 25 m long boat that allows for coastal sampling, mainly of shallow areas. Its autonomy is 22 days, and it can hold 13 people, including researchers, technicians, and crew.

It is equipped with a SIMRAD EK60 echosounder with two 38 and 120 kHz transducers embedded in the boat's hull. It has winches to work with demersal or pelagic nets (Table 3). Since 2008, this vessel has carried out uninterrupted acoustic surveys and maintains a busy schedule of annual surveys, mainly assessing small pelagic fish, Pacific hake, and giant squid, and some extraordinary surveys that have been carried out in northwestern Mexico (e.g., of totoaba). This vessel has provided the longest time series of acoustic data of small pelagic species in Mexico (2009–2023).

Table 3. Platforms and equipment available for acoustic surveys in Mexico.

Platforms	Institution	Echosounder System	Characteristics
RV BIP XI Autonomy: 22 days	IMIPAS	EK60	Split-beam transducer ES38-12 (resonant frequency: 38 kHz; beamwidth: 11.94°) Split-beam transducer ES120-7C (resonant frequency: 120 kHz; beamwidth: 7.61°) Source: manuel.nevarez@imipas.gob.mx
RV Dr. Jorge Carranza Fraser Autonomy: 60 days	IMIPAS	EK60	Split-beam transducer ES18 312 (resonant frequency: 18 kHz; beamwidth: 10.53°) Split-beam transducer ES38-B 312 (resonant frequency: 38 kHz; beamwidth: 6.79°) Split-beam transducer ES70-7C 312-204154 (side) (resonant frequency: 70 kHz; beamwidth: 7.12°) Split-beam transducer ES70-7C 312-204154 (resonant frequency: 70 kHz; beamwidth: 7.12°) Split-beam transducer ES120-7C 312-204022 (resonant frequency: 120 kHz; beamwidth: 7.29°) Split-beam transducer ES200-7C 312-200841 (side and down) (resonant frequency: 200 kHz; beamwidth: 7.45°) Source: https://www.gob.mx/agricultura/colima/articulos/conoce-nuestro-buque-de-investigacion-dr-jorge-carranza-fraser?idiom=es, accessed on 4 September 2024
RV El Puma Autonomy: 30 days	UNAM	EK60	Frequency: 38 kHz Frequency: 120 kHz Source: https://buques.cic.unam.mx/el-puma/equipo-puma/, accessed on 4 September 2024
RV Justo Sierra Autonomy: 30 days	UNAM	EK80	Wide-band split-beam transducer ES38-7 Resonant frequency: 38 kHz Beamwidth: 7° Source: https://buques.cic.unam.mx/justo-sierra/equipo-js/, accessed on 4 September 2024
Swing steel arm	CICIMAR	EK80	Combi transducer ES38-18/200-18C Nominal frequency Low: 38 kHz High: 200 kHz Beamwidth: 18° Source: hvillalo@ipn.mx
Swing steel arm	IMIPAS	EK80	Wide-band split-beam transducer ES38-7 Resonant frequency: 38 kHz Beamwidth: 7.5° Source: manuel.nevarez@imipas.gob.mx

RV Dr. Jorge Carranza Fraser is almost 60 m long. It was financed by the Inter-American Development Bank and is undoubtedly the most important investment in Mexico for acoustics research. The acoustic system comprises seven transducers at the frequencies of 18, 38, 70, 120, and 200 kHz and two side frequencies of 70 kHz, coupled to a SIMRAD EK60 echosounder. It is built to be autonomous for 60 days, which allows it to reach the

Exclusive Economic Zone, and can house 22 researchers and technicians. Its programming includes multipurpose campaigns for several species, mainly small pelagic species, Pacific hake, and squid. This modern vessel has advantages over older ships: a quiet diesel–electric engine, a liftable keel to prevent bubble reverberation, and better modern fishing facilities; these features provide remarkable results and spatial coverage.

Since 1980, The National Autonomous University of Mexico (UNAM) has managed RV El Puma. Although this vessel was initially used for acoustic surveys, it is currently in limited use for this purpose. RV El Puma has an autonomy of 30 days and carries a SIMRAD EK60 echosounder with a 38 kHz transducer that has been used mainly to investigate plankton and pelagic organisms such as squids and small pelagic fish. The EK60 echosounder and its ER60 data acquisition software are no longer manufactured. The SIMRAD EK80 system has taken its place; it is a more advanced system that has broadband capability and a significant improvement in the signal-to-noise ratio. INAPESCA and CICIMAR acquired this equipment, which has been installed in and uninstalled from different platforms, such as small boats, via a swing steel arm. These devices allow autonomy to carry out sampling in different regions, and the vessel can even reach continental waterbodies, where research with this technology has not yet been carried out.

Other vessels carry out marine research in Mexico, including the Alpha Helix Oceanographic Vessel, which is the property of the Center for Scientific Research and Higher Education at Ensenada (CICESE). Although it has carried out extensive hydrographic and planktonic campaigns, it does not integrate biological acoustics as a research goal; therefore, it does not have this kind of echosounder installed. RV Justo Sierra, which is also the property of UNAM, has recently installed an EK80 multifrequency echosounder (Table 3). The development of acoustic instrumentation is constant, and the miniaturization of echosounders has triggered the development of instruments whose potential is expanding every decade, in a stage that some have called the robot revolution [142].

Although it is not the aim of this review to enumerate global advances on this topic, there are promising developments of alternative platforms that provide opportunities to address new questions with fewer investments and operating challenges than a ship. For those institutions that do not own ships, unmanned surface vehicles (USVs) have proved to be operationally efficient and have the potential to be incorporated into research programs [143]. According to Eriksen et al. [144], there are advantages to coupling echosounders: This approach provides a depth profile of plankton (macrozooplankton and mesozooplankton) and fish aggregations in water masses that can be coupled to hydrological–oceanographic stations that are already being used on Mexican vessels. Although these developments are different from the "traditional" approach, they have the potential to increase the information available and can be incorporated by institutions interested in expanding their interdisciplinary research programs. Given the limited human resources of some Mexican institutions, investing in instrumentation to develop this research goal is unrealistic in the short term. However, capitalizing on investments through scientific collaboration schemes to maximize the benefit of the campaigns—and thus obtain more and better information and biological data—could be assumed to be a medium-term goal.

7.2. Investment in Human Resources

Fisheries acoustics is an applied science that requires a multidisciplinary approach between physics and engineering that is framed in a biological context. Acoustic surveys should be conducted in the presence of at least one scientist who is experienced in fisheries acoustics and is familiar with the target species in the surveyed area. This scientist can guide the fishery and recognize the potential of complementary data. In Mexico, there are few people with experience and formal training in the operation of acoustic equipment. Indeed, there are not enough human resources for the number of acoustic campaigns that are currently being carried out inside and outside the country with Mexican investment. A lot of acoustic data are piling up, and there is a failure to recognize the potential that this information has for understanding the assemblages that exist between zooplankton, ichthyoplankton, and mesopelagic organisms beyond the assessment of the most commercially important species (e.g., small pelagic fish, jumbo squid, and Pacific hake). Future challenges include developing a strategy to build long-term capabilities to understand and use acoustic techniques through undergraduate or postgraduate education. This approach should include updates from national experts, developing skills to capitalize on existing investments, and maximizing the applicability of acoustic techniques. Engaging students in solving real-world problems and linking the needs of the country with the needs of universities would provide them with direct research opportunities and experience. This collaboration could lead to the development of new methods that respond to the demands of the complex and contrasting ecosystems exhibited in Mexican waters, such as mixed aggregations and species assemblages, for which the acoustic characterization of locally distributed species is unknown.

Through its academic unit, Centro Interdisciplinario de Ciencias Marinas (CICIMAR), National Polytechnic Institute (IPN) has developed a postgraduate program in which the topic of fisheries acoustics has recently expanded. Although the working group is limited, it has had collaborations, mainly with INAPESCA, to assess marine resources. In this sense, there are experiences from IPN that provide advice and develop research focused on marine acoustics. While these efforts satisfy some demands, a single institution cannot solve all problems. Promoting exchanges between North American and European universities, which have solid postgraduate courses and consolidated working groups, would be a useful choice. The advantages of marine acoustics have been recognized by stakeholders in Mexican institutions with a focus on fishery management as necessary in contributing to the sustainability of harvested stocks. Estimates based on fishery-independent data are valuable because they can be compared with stock assessments supported by fishery-dependent data. This comparison could allow researchers to test hypotheses regarding stock conditions and to provide recommendations to enhance the management instruments of the Mexican government.

8. Conclusions

Improving fishery management in a changing environment is a global challenge. Mexico must face this challenge with solid scientific tools and modern assessment methods to streamline recommendations and the decision-making process. This approach would comply with the objectives of sustainable development and contribute to improving the quality of life of coastal populations while generating a steady supply of fish protein.

The successful application of fishery acoustics in Mexico has been documented for the Pacific sardine (*Sardinops sagax*) fishery in the Gulf of California. Acoustic data have been useful for providing biomass estimates and indices of relative abundance; these estimations are usually contrasted with biomass estimated from statistical models expressing the population dynamics for this species; under this scenario, it is possible to obtain information from fishery-independent data (an acoustic survey) and fishery-dependent data [145]. The acoustic outcomes have been used to support the management of the Pacific sardine fishery; this endeavor is included in the legal Mexican guidelines through the Fishery National Chart [19], the Fishery Management Plan [146], and the Mexican Official Standard [147]. Additionally, the scientific information reported using acoustic data has been used to obtain international certification of this fishery from the Marine Stewardship Council (https://fisheries.msc.org/en/fisheries/small-pelagics-fishery-in-sonora-gulf-of-california/@@view; accessed on 5 July 2024). The Pacific thread herring (*Ophistonema libertate*) fishery in the south zone from the Gulf of California has also received this certification (https://fisheries.msc.org/en/fisheries/southern-gulf-of-california-thread-herring/@@view; accessed on 5 July 2024).

Although fisheries acoustics still present technical and scientific challenges, the field is expanding to meet the growing demands of modern ecosystemic and multidisciplinary marine research. Traditional ship-mounted sensors remain prevalent, but emerging marine technologies such as autonomous underwater vehicles (AUVs), remotely operated vehicles

(ROVs), unmanned surface vehicles (USVs), moored inverted echosounders, and compact high-capacity echosounders have significant potential for integration into existing fishery surveys and research projects [148]. These new platforms not only increase the diversity and volume of data but also allow operations during periods when surveys are impractical, thereby expanding both the spatial and temporal coverage. Together, these advances could make long-term acoustic measurements of fish abundance more accessible and are likely to see broader application in the future [10,149]. For Mexico, these platforms could potentially enable the collection of biomass data for other marine resources that are currently not monitored.

The current state of development of acoustic instruments has required considerable economic investments and interdisciplinary groups of scientists and engineers who have led this discipline toward constant improvement and diversification. The countries maintaining a leading role in the use, development, and innovation of acoustic instruments have implemented a long-term vision that includes collaboration among their institutions. This approach will allow them to continue to implement strategies for the management and conservation of marine resources and to improve their position in the face of emerging challenges that arise from current problems.

For developing countries such as Mexico, fishing is a primary source of food and livelihood for coastal communities. The delay in implementing fisheries acoustics in countries with these characteristics is an example of failed technology transfer and contributes to the underrepresentation of information on tropical and subtropical seas, which may represent a gap in the understanding of the response of marine communities to global changes. For the past decade, Mexico has made a sustained effort to implement acoustic techniques to provide a time series of standardized information. However, at present their acoustic data are underutilized. Because the current global approach to fishery management is ecosystem-based, ecological time series become more informative when they are longer. In Mexico, survey research usually evaluates a single species, so the surveyed area, the fishing gear, and the type of vessel have always been used for surveying the most abundant species. This type of approach is usually the only one that a small RV can address due to its space or autonomy limitations. The Mexican deep-draft fleet (RV Dr. Jorge Carranza Fraser and RV El Puma) can change this situation by participating in campaigns that improve the understanding of a wide range of ecological subjects so that an ecosystemic approach can be implemented from marine acoustic data.

The Mexican scientific community maintains a good understanding of how technology and science can make a positive contribution to fishery management without "ready-to-use" solutions. The gap between commercial and scientific development is becoming smaller; acoustic tools developed for commercial purposes are being adopted by scientists, and scientific developments are being used to produce commercial tools. These changes represent opportunities that can be taken advantage of to shorten the path. Consistent long-term studies will enhance the response to emerging problems such as climate change and the redistribution of species, so it is necessary to continue to support research programs that have been maintained in the long term in Mexico. For Mexico, to at least maintain the current infrastructure for acoustic survey research, there is a need to increase the budget for the development of monitoring programs that collect ecosystem indicator data, train human resources, and encourage peer review of the information generated and reported in gray literature.

One of the purposes of this review is to promote the use of acoustics to study aquatic ecosystems, especially considering that other methodological developments or currents of thought have failed to become established in the Mexican scientific community and have simply disappeared from the country. To ensure that the initiatives mentioned in this work prevail, evolve, and transcend, the discussion on the contributions of acoustics to fishery management, MPAs, and the ecosystemic approach must be built, defended, and promoted among the community of Mexican fishery scientists.

Author Contributions: Conceptualization, V.E.G.-M. and E.M.-B.; methodology, V.E.G.-M. and E.M.-B.; validation, H.V. and M.O.N.-M.; investigation, V.E.G.-M.; writing—original draft preparation, V.E.G.-M. and E.M.-B.; writing—review and editing, V.E.G.-M. and E.M.-B.; visualization, V.E.G.-M. and E.M.-B.; supervision, H.V. and M.O.N.-M. All authors have read and agreed to the published version of the manuscript.

Funding: The authors thank CONAHCYT for the Ph.D. scholarship (number 782714).

Institutional Review Board Statement: Not applicable.

Informed Consent Statement: Not applicable.

Data Availability Statement: Not applicable.

Acknowledgments: The authors wish to especially thank the technicians and researchers from IMIPAS/CRIAP Guaymas for their valuable contributions to the development of multiple acoustic campaigns carried out over the recent decade. The opinions expressed in this publication are directly attributable to the authors and do not represent the opinions or policies of the Mexican institutions. Any use of trademarks, commercial products, or company names is for descriptive purposes only.

Conflicts of Interest: The authors declare no conflicts of interest.

References

1. Dornan, T.; Fielding, S.; Saunders, R.A.; Genner, M.J. Large Mesopelagic Fish Biomass in the Southern Ocean Resolved by Acoustic Properties. *Proc. R. Soc. B* **2021**, *289*, 20211781. [CrossRef] [PubMed]
2. Stenevik, E.K.; Vølstad, J.H.; Høines, Å.; Aanes, S.; Óskarsson, G.J.; Jacobsen, J.A.; Tangen, Ø. Precision in Estimates of Density and Biomass of Norwegian Spring-Spawning Herring Based on Acoustic Surveys. *Mar. Biol. Res.* **2015**, *11*, 449–461. [CrossRef]
3. Wassermann, S.; Johnson, M.P. The Potential to Improve the Sustainability of Pelagic Fisheries in the Northeast Atlantic by Incorporating Individual Fish Behavior Into Acoustic Sampling. *Front. Mar. Sci.* **2020**, *7*, 357. [CrossRef]
4. David, V.; Mouget, A.; Perrot, Y.; Le Goff, L.; Thiriet, P.; Diogoul, N.; Feunteun, E.; Acou, A.; Brehmer, P. Insights from a Multibeam Echosounder to Survey Pelagic Fish Shoals and Their Spatio-Temporal Distribution in Ultra-Shallow Waters. *Estuar. Coast. Shelf Sci.* **2022**, *264*, 107705. [CrossRef]
5. Domokos, R. On the Development of Acoustic Descriptors for Semi-Demersal Fish Identification to Support Monitoring Stocks. *ICES J. Mar. Sci.* **2021**, *78*, 1117–1130. [CrossRef]
6. Shao, H.; Kiyomoto, S.; Kadota, T.; Nakagawa, M.; Yamanaka, H.; Kawauchi, Y.; Minami, K.; Miyashita, K. Assessment of Spatio-Temporal Variations of Macroalgal Canopies and Fish Schools before and after Coastal Desertification Using Acoustic Methods. *Hydrobiologia* **2023**, *851*, 1891–1906. [CrossRef]
7. Lu, Z.; Mukai, T.; Fujimori, Y.; Lida, K. Estimating the Sampling Efficiencies of a Framed Midwater Trawl and Ring Net for Zooplankton Using an Acoustic Method Net for Zooplankton Using an Acoustic Method. *J. Mar. Sci. Technol.* **2021**, *29*, 3. [CrossRef]
8. Zhang, J.; Chen, Z.-Z.; Chen, G.-B.; Zhang, P.; Qiu, Y.-S.; Yao, Z. Hydroacoustic Studies on the Commercially Important Squid *Sthenoteuthis oualaniensis* in the South China Sea. *Fish. Res.* **2015**, *169*, 45–51. [CrossRef]
9. Yang, H.; Cheng, J.; Tang, T.; Chen, J.; Li, G. Acoustic Target Strength of Jellyfish, *Nemopilema nomurai*, Measured at Multi-Frequency and Multi-Orientation. *J. Appl. Ichthyol.* **2023**, *2023*, 6650863. [CrossRef]
10. Trenkel, V.M.; Handegard, N.O.; Weber, T.C. Observing the Ocean Interior in Support of Integrated Management. *ICES J. Mar. Sci.* **2016**, *73*, 1947–1954. [CrossRef]
11. Benoit-Bird, K.J.; Lawson, G.L. Ecological Insights from Pelagic Habitats Acquired Using Active Acoustic Techniques. *Annu. Rev. Mar. Sci.* **2016**, *8*, 463–490. [CrossRef] [PubMed]
12. Brown, C.J.; Smith, S.J.; Lawton, P.; Anderson, J.T. Benthic Habitat Mapping: A Review of Progress towards Improved Understanding of the Spatial Ecology of the Seafloor Using Acoustic Techniques. *Estuar. Coast. Shelf Sci.* **2011**, *92*, 502–520. [CrossRef]
13. Fernandes, P.G.; Gerlotto, F.; Holliday, D.V.; Nakken, O.; Simmonds, E.J. Acoustic Applications in Fisheries Science: The ICES Contribution. *ICES Mar. Sci. Symposia* **2002**, *215*, 483–492. [CrossRef]
14. Alverson, D.L. *FAO Study Tour in USSR*; US Fish and Wildlife Service: Washington, DC, USA, 1967; Sep No. 805.
15. Bhupinder Singh, D.; Andrews, F. *A Literature Survey on the Subject of the Use of Acoustics in Fish Catching and Fish Study*; Institute of Ocean Engineering, School of Engineering and Architecture, The Catholic University of America: Washington, DC, USA, 1971.
16. Dragesund, O.; Midttun, L. Development of Acoustic Techniques in Norway for Fisheries Research and Commercial Fishing. *Proc. R. Soc. Edinb. Sect. B Biol.* **1972**, *73*, 429–435. [CrossRef]
17. Melcer-Zaiane, J.; García-Franco, W.; Mondragon-Corona, E.; Cota-Villavicencia, A. Estimación de Biomasa y Distribución de Peces Pelágicos Con Métodos Hidroacústicos En La Corriente de California Frente a La Costa Occidental de Baja California. In Proceedings of the Primer Simposium Nacional de Recursos Pesqueros Masivos de México, Ensenada, Mexico, 28–30 September 1976.

18. Cotero-Altamirano, C.E.; Green-Ruiz, Y. *Spawning Biomass of the Northern Anchovy (Engraulis mordax) in the Gulf of California during 1991*; CalCOFI Reo: La Jolla, CA, USA, 1997; Volume 38.
19. Diario Oficial de la Federación. *Acuerdo Mediante el Cual Se da a Conocer la Actualización de la Carta Nacional Pesquera*; Diario Oficial de la Federación: Cuauhtémoc, Mexico, 2021.
20. Korneliussen, R.J.; Heggelund, Y.; Macaulay, G.; Patel, D.; Johnsen, E.; Eliassen, I.K. Acoustic Identification of Marine Species Using a Feature Library. *Methods Oceanogr.* **2016**, *17*, 187–205. [CrossRef]
21. Knudsen, H.P. The Bergen Echo Integrator: An Introduction. *ICES J. Mar. Sci.* **1990**, *47*, 167–174. [CrossRef]
22. Perrot, Y.; Brehmer, P.; Habasque, J.; Roudaut, G.; Behagle, N.; Sarré, A.; Lebourges-Dhaussy, A. Matecho: An Open-Source Tool for Processing Fisheries Acoustics Data. *Acoust. Aust.* **2018**, *46*, 241–248. [CrossRef]
23. Ballón, M.; Bertrand, A.; Lebourges-Dhaussy, A.; Gutiérrez, M.; Ayón, P.; Grados, D.; Gerlotto, F. Is There Enough Zooplankton to Feed Forage Fish Populations off Peru? An Acoustic (Positive) Answer. *Prog. Oceanogr.* **2011**, *91*, 360–381. [CrossRef]
24. Ladroit, Y.; Escobar-Flores, P.C.; Schimel, A.C.G.; O'Driscoll, R.L. ESP3: An Open-Source Software for the Quantitative Processing of Hydro-Acoustic Data. *Softwarex* **2020**, *12*, 100581. [CrossRef]
25. Johnsen, E.; Totland, A.; Skålevik, A.; Holmin, A.J.; Dingsør, G.E.; Fuglebakk, E.; Handegard, N.O. StoX: An Open Source Software for Marine Survey Analyses. *MEE* **2019**, *10*, 1523–1528. [CrossRef]
26. Villar, S.; Madirolas, A.; Cabreira, A.G.; Rozenfeld, A.; Acosta, G.G. ECOPAMPA: A New Tool for Automatic Fish Schools Detection and Assessment from Echo Data. *Heliyon* **2021**, *7*, e05906. [CrossRef] [PubMed]
27. Lee, W.-J.; Mayorga, E.; Setiawan, L.; Majeed, I.; Nguyen, K.; Staneva, V. Echopype: A Python Library for Interoperable and Scalable Processing of Water Column Sonar Data for Biological Information. *arXiv* **2021**, arXiv:2111.00187. [CrossRef]
28. ICES. *Working Group on Fisheries Acoustics, Science and Technology (WGFAST)*; Scientific Reports; ICES: Copenhagen, Denmark, 2020; Volume 2, p. 18. [CrossRef]
29. Wall, C.C.; Towler, R.; Anderson, C.; Cutter, R.; Jech, J.M. PyEcholab: An Open-Source, Python-Based Toolkit to Analyze Water-Column Echosounder Data. *J. Acoust. Soc. Am.* **2018**, *144*, 1778. [CrossRef]
30. R Core Team. *R: A Language and Environment for Statistical Computing*; R Foundation for Statistical Computing: Vienna, Austria, 2022.
31. Villalobos, H. *Echogram: Echogram Visualisation and Analysis*; R Package, Version 0.1.2; R Foundation for Statistical Computing: Vienna, Austria, 2022. Available online: https://CRAN.R-project.org/package=echogram (accessed on 4 September 2024).
32. Weill, A.; Scalabrin, C.; Diner, N. MOVIES-B: An Acoustic Detection Description Software. Application to Shoal Species Classification. *Aquat. Living Resour.* **1993**, *6*, 255–267. [CrossRef]
33. Reid, D.G. Report on Echo Trace Classification. *ICES Coop. Res. Rep.* **2000**, *238*, 115. [CrossRef]
34. Korneliussen, R.J.; Ona, E.; Eliassen, I.; Heggelund, Y.; Patel, R.; Godø, O.R.; Giertsen, C.; Patel, D.; Nornes, E.; Bekkvik, T.; et al. The Large Scale Survey System—LSSS. In Proceedings of the 29th Scandinavian Symposium on Physical Acoustics, Ustaosetm, Norway, 29 January–1 February 2006.
35. Berger, L.; Durand, C.; Marchalot, C.; Diner, N. *Movies+ User Manual Version 4.3*; Ifremer: Plouzané, France, 2005.
36. Korneliussen, R.J. *Advances in Bergen Echo Integrator*; ICES Cooperative Report; ICES: Copenhagen, Denmark, 1993; p. 12.
37. Simmonds, E.J.; Williamson, N.J.; Gerlotto, F.; Aglen, A. *Acoustic Survey Design and Analysis Procedure: A Comprehensive Review of Current Practice*; ICES Cooperative Report; ICES: Copenhagen, Denmark, 1992; Volume 187.
38. Walline, P.D. Geostatistical Simulations of Eastern Bering Sea Walleye Pollock Spatial Distributions, to Estimate Sampling Precision. *ICES J. Mar. Sci.* **2007**, *64*, 559–569. [CrossRef]
39. Kimura, D.; Lemberg, N.A. Variability of Line Intercept Density Estimates (A Simulation Study of the Variance of Hydroacoustic Biomass Estimates). *Can. J. Fish. Aquat. Sci.* **1981**, *38*, 1141–1152. [CrossRef]
40. Kimura, D.K.; Somerton, D.A. Review of Statistical Aspects of Survey Sampling for Marine Fisheries. *Rev. Fish. Sci.* **2007**, *14*, 245–283. [CrossRef]
41. López-Serrano, A.; Villalobos-Ortíz, H.; Nevárez-Martínez, M.O. A Probabilistic Procedure for Estimating an Optimal Echointegration Threshold Using the Expectation-Maximisation Algorithm. *Aquat. Living Resour.* **2018**, *31*, 12. [CrossRef]
42. Enciso-Enciso, C.; Nevárez-Martínez, M.O.; Sánchez-Cárdenas, R.; Salcido-Guevara, L.A.; Minte-Vera, C.; Marín-Enriquez, E.; Hernández-Rivas, M. Assessment and Management of the Temperate Stock of Pacific Sardine (*Sardinops sagax*) in the South of California Current System. *Reg. Stud. Mar. Sci.* **2023**, *62*, 102972. [CrossRef]
43. Rubio-Rodríguez, U.; Villalobos, H.; Nevárez-Martínez, M.O. Acoustic Observations of the Vertical Distribution and Latitudinal Range of Small Pelagic Fish Schools in the Midriff Islands Region, Gulf of California, Mexico. *Lat. Am. J. Aquat. Res.* **2018**, *46*, 989–1000. [CrossRef]
44. Fréon, P.; Misund, O.A. *Dynamics of Pelagic Fish Distribution and Behaviour: Effects on Fisheries and Stock Assessment*; Fishing New Books, Blackwell Science Ltd.: Oxford, UK, 1999; p. 348.
45. Morales-Bojórquez, E.; Nevárez-Martínez, M.O.; García-Alberto, G.; Villalobos, H.; Aguirre-Villaseñor, H.; Larios-Castro, E.; González-Peláez, S.S.; Arizmendi-Rodríguez, D.I.; Martínez-Zavala, M.A. Interaction Between Marine Fauna and the Small Pelagic Fishery in the Coastal Environment of the Gulf of California, Mexico. *Front. Mar. Sci.* **2021**, *8*, 669176. [CrossRef]
46. Leonori, I.; Ticina, V.; Giannoulaki, M.; Hattab, T.; Iglesias, M.; Bonanno, A.; Costantini, I.; Canduci, G.; Machias, A.; Ventero, A.; et al. History of Hydroacoustic Surveys of Small Pelagic Fish Species in the European Mediterranean Sea. *Mediterr. Mar. Sci.* **2021**, *22*, 751–768. [CrossRef]

47. Salthaug, A.; Stenevik, K.E.; Vatnehol, S.; Anthonypillai, V.; Slotte, A. *Distribution and Abundance of Norwegian Springspawning Herring during the Spawning Season in 2021*; Survey Report; Institute of Marine Research: Bergen, Norway, 2021; ISSN 15036294.
48. Cutter, G.R.; Demer, D.A. *California Current Ecosystem Survey 2006 Acoustic Cruise Reports for NOAA FSV Oscar Dyson and NOAA FRV David Starr Jordan*; NOAA Technical Memo; NOAA-SWFSC-415: 98; U.S. Department of Commerce: Washington, DC, USA, 2008.
49. Demer, D.A.; Hewitt, R.P. Bias in Acoustic Biomass Estimates of *Euphausia superba* Due to Diel Vertical Migration. *Deep Sea Res. I Oceanogr. Res. Pap.* **1995**, *42*, 455–475. [CrossRef]
50. Doray, M.; Badts, V.; Masse, J.; Huret, M.; Doremus, G.; Ptetitgas, P. *Manual of Fisheries Survey Protocols 2014 PELGAS Surveys (PELagiques GAScogne)*; RBE/EMH 2014-01; Ifremer: Plouzane, France, 2014. [CrossRef]
51. Ohshimo, S. Spatial Distribution and Biomass of Pelagic Fish in the East China Sea in Summer, Based on Acoustic Surveys from 1997 to 2001. *Fish. Sci.* **2004**, *70*, 389–400. [CrossRef]
52. Koslow, J. The Role of Acoustics in Ecosystem-Based Fishery Management. *ICES J. Mar. Sci.* **2009**, *66*, 966–973. [CrossRef]
53. Polovina, J.J. Model of a Coral Reef Ecosystem. *Coral Reefs* **1984**, *3*, 1–11. [CrossRef]
54. Pauly, D.; Christensen, C.V.; Walters, C.J. Ecopath, Ecosim, and Ecospace as Tools for Evaluating Ecosystem Impact of Fisheries. *ICES J. Mar. Sci.* **2000**, *57*, 697–706. [CrossRef]
55. Krebs, J.C.; Danell, K. Terrestrial Trophic Dynamics in the Canadian Arctic. *Can. J. Zool.* **2003**, *81*, 827–843. [CrossRef]
56. Gascuel, D.; Pauly, D. EcoTroph: Modelling Marine Ecosystem Functioning and Impact of Fishing. *Ecol. Model.* **2009**, *220*, 2885–2898. [CrossRef]
57. Christensen, V.; Walters, C.; Pauly, D.; Forrest, R. *Ecopath with Ecosim Version 6: User Guide*; Fisheries Centre, University of British Columbia: Vancouver, BC, Canada; ICLARM: Penang, Malaysia, 2008; p. 130.
58. Rose, K.A.; Allen, J.I. End-To-End Models for the Analysis of Marine Ecosystems: Challenges, Issues, and Next Steps. *Mar. Coast. Fish.* **2010**, *2*, 115–130. [CrossRef]
59. Handegard, N.O.; Buisson, L.; Brehmer, P.; Chalmers, S.J.; de Robertis, A.; Huse, G.; Kloser, R.; Macaulay, G.; Maury, O.; Ressler, P.H.; et al. Towards an Acoustic-Based Coupled Observation and Modelling System for Monitoring and Predicting Ecosystem Dynamics of the Open Ocean. *Fish Fish.* **2013**, *14*, 605–615. [CrossRef]
60. Del Monte-Luna, P.; Arreguín-Sánchez, F.; Lluch-Belda, D. Marine ecosystem analyses in the Gulf of Ulloa, Mexico: BAC meets Ecopath. In *INCOFISH Ecosystem Models: Transiting from Ecopath to Ecospace*; Le Quesne, W., Arreguín-Sánchez, F., Heymans, S., Eds.; Fisheries Centre Research Reports; University of British Columbia: Vancouver, BC, Canada, 2007; Volume 15, pp. 114–133.
61. Godø, O.R.; Handegard, N.O.; Browman, H.I.; Macaulay, G.; Kaartvedt, S.; Giske, J.; Ona, E.; Huse, G.; Johnsen, E. Marine Ecosystem Acoustics (MEA): Quantifying Processes in the Sea at the Spatio-Temporal Scales on Which They Occur. *ICES J. Mar. Sci.* **2014**, *71*, 2357–2369. [CrossRef]
62. Bertrand, A.; Josse, E.; Bach, P.; Dagorn, L. Acoustics for Ecosystem Research: Lessons and Perspectives from a Scientific Programme Focusing on Tuna-Environment Relationships. *Aquat. Living Resour.* **2003**, *16*, 197–203. [CrossRef]
63. Lazzari, M.A.; Tupper, B. Importance of Shallow Water Habitats for Demersal Fishes and Decapod Crustaceans in Penobscot Bay, Maine. *Environ. Biol. Fishes* **2002**, *63*, 57–66. [CrossRef]
64. Collings, W.; McConnaughey, R.A. Acoustic Classification of the Sea Floor to Address Essential Fish Habitat and Marine Protected Area Requirements. In Proceedings of the Canadian Hydrographic Conference, Victoria, BC, Canada, 10–12 March 1998; pp. 369–377.
65. Kloser, R.J.; Penrose, J.D.; Butler, A.J. Multi-Beam Backscatter Measurements Used to Infer Seabed Habitats. *Cont. Shelf Res.* **2010**, *30*, 1772–1782. [CrossRef]
66. Lamarche, G. Quantitative Characterization of Seafloor Substrate and Bedforms Using Advanced Processing of Multibeam Backscatter—Application to Cook Strait, New Zealand. *Cont. Shelf Res.* **2011**, *31*, S93–S109. [CrossRef]
67. Cutter, G.R.; Demer, D.A. Seabed Classification Using Surface Backscattering Strength versus Acoustic Frequency and Incidence Angle Measured with Vertical, Split-Beam Echosounders. *ICES J. Mar. Sci.* **2014**, *71*, 882–894. [CrossRef]
68. Siwabessy, J.W.; Tseng, Y.-T.; Gavrilov, A.N. Seabed Habitat Mapping in Coastal Waters Using a Normal Incident Acoustic Technique. In Proceedings of the Acoustics, Gold Coast, Australia, 3–5 November 2004.
69. Gastauer, S.; Scoulding, B.; Parsons, M. An Unsupervised Acoustic Description of Fish Schools and the Seabed in Three Fishing Regions within the Northern Demersal Scalefish Fishery (NDSF, Western Australia). *Acoust. Aust.* **2017**, *45*, 363–380. [CrossRef]
70. Pitcher, T.J.; Kalikoski, D.; Short, K.; Varkey, D.; Pramod, G. An Evaluation of Progress in Implementing Ecosystem-Based Management of Fisheries in 33 Countries. *Mar. Policy* **2009**, *33*, 223–232. [CrossRef]
71. Morales-Zarate, M.V.; Arreguín-Sánchez, F.; López-Martínez, J.; Lluch-Cota, S.E. Ecosystem Trophic Structure and Energy Flux in the Northern Gulf of California, México. *Ecol. Model.* **2004**, *174*, 331–345. [CrossRef]
72. Arreguín-Sánchez, F.; Arcos, E.; Chávez, E.A. Flows of Biomass and Structure in an Exploited Benthic Ecosystem in the Gulf of California, Mexico. *Ecol. Model.* **2002**, *156*, 167–183. [CrossRef]
73. Rosas-Luis, R.; Salinas-Zavala, C.A.; Koch, V.; Del Monte-Luna, P.; Morales-Zarate, M.V. Importance of Jumbo Squid *Dosidicus gigas* (Orbigny, 1835) in the Pelagic Ecosystem of the Central Gulf of California. *Ecol. Model.* **2008**, *218*, 149–161. [CrossRef]
74. Morales-Zarate, M.V.; Lopez-Ramírez, J.A.; Salinas-Zavala, C.A. Loggerhead Marine Turtle (*Caretta caretta*) Ecological Facts from a Trophic Relationship Model in a Hot Spot Fishery Area: Gulf of Ulloa, Mexico. *Ecol. Model.* **2021**, *439*, 109327. [CrossRef]

75. Gómez- Gutiérrez, J.; Robinson, C.J. Tidal Current Transport of Epibenthic Swarms of the Euphausiid *Nyctiphanes simplex* in a Shallow, Subtropical Bay on Baja California Peninsula, México. *Mar. Ecol. Progr. Ser.* **2006**, *320*, 215–231. [CrossRef]
76. Robinson, C.J.; Anislado, V.; Ramos, P. Shoaling Fish and Red Crab Behaviour Related to Tidal Variations in Bahía Magdalena, México. *Deep Sea Res. II Top. Stud. Oceanogr.* **2004**, *51*, 767–783. [CrossRef]
77. Ursella, L.; Pensieri, S.; Pallás-Sanz, E.; Herzka, S.Z.; Bozzano, R.; Tenreiro, M.; Cardin, V.; Candela, J.; Sheinbaum, J. Diel, Lunar and Seasonal Vertical Migration in the Deep Western Gulf of Mexico Evidenced from a Long-Term Data Series of Acoustic Backscatter. *Prog. Oceanogr.* **2021**, *195*, 102562. [CrossRef]
78. Portner, E.J.; Benoit-Bird, K.J.; Hazen, E.L.; Waluk, C.M.; Robinson, C.J.; Gómez-Gutiérrez, J.; Gilly, W.F. Decline and Recovery of Pelagic Acoustic Backscatter Following El Niño Events in the Gulf of California, Mexico. *Prog. Oceanogr.* **2022**, *206*, 102823. [CrossRef]
79. Sarmiento-Lezcano, A.N.; Busquets-Vass, G.; Rubio-Rodríguez, U.; Pilar Olivar, M.; Peña, M.; Medina-Suárez, I.; González-Rodríguez, E.; Gómez- Gutiérrez, J.; Robinson, C.J.; Hernández-León, S. Active Flux Seasonality of the Small Dominant Migratory Crustaceans and Mesopelagic Fishes in the Gulf of California during June and October. *Prog. Oceanogr.* **2022**, *208*, 102894. [CrossRef]
80. Crain, C.M.; Halpern, B.S.; Beck, M.W.; Kappel, C.V. Understanding and Managing Human Threats to the Coastal Marine Environment. *Ann. N. Y. Acad. Sci.* **2009**, *1162*, 39–62. [CrossRef]
81. Ortíz-Lozano, L.; Olivera-Vázquez, L.; Espejel, I. Legal Protection of Ecosystem Services Provided by Marine Protected Areas in Mexico. *Ocean Coast. Manag.* **2017**, *138*, 101–110. [CrossRef]
82. Rife, A.N.; Erisman, B.; Sanchez, A.; Aburto-Oropeza, O. When Good Intentions Are Not Enough... Insights on Networks of "Paper Park" Marine Protected Areas. *Conserv. Lett.* **2013**, *6*, 200–212. [CrossRef]
83. Aburto-Oropeza, O.; Erisman, B.; Galland, G.R.; Mascareñas-Osorio, I.; Sala, E.; Ezcurra, E. Large Recovery of Fish Biomass in a No-Take Marine Reserve. *PLoS ONE* **2011**, *6*, e23601. [CrossRef] [PubMed]
84. Kelleher, G.; Phillips, A. *Guidelines for Marine Protected Areas*; IUCN: Gland, Switzerland; Cambridge, UK, 1999; pp. 24–107.
85. Bloomfield, H.J.; Sweeting, C.; Mill, A.C.; Stead, S.M.; Polunin, N.V.C. No-Trawl Area Impacts: Perceptions, Compliance and Fish Abundances. *Environ. Conserv.* **2012**, *9*, 237–247. [CrossRef]
86. Hill, N.A.; Barrett, N.; Lawrence, E.; Hulls, J.; Dambacher, J.M.; Nichol, S.; Williams, A.; Hayes, K.R. Quantifying Fish Assemblages in Large, Offshore Marine Protected Areas: An Australian Case Study. *PLoS ONE* **2014**, *9*, e110831. [CrossRef]
87. Starr, R.M.; Wendt, D.E.; Barnes, C.L.; Marks, C.I.; Malone, D.; Waltz, G.; Schmidt, K.T.; Chiu, J.; Launer, A.L.; Hall, N.C.; et al. Variation in Responses of Fishes across Multiple Reserves within a Network of Marine Protected Areas in Temperate Waters. *PLoS ONE* **2015**, *10*, e0118502. [CrossRef]
88. Egerton, J.P.; Turner, J.; LeVay, L.; Mascareñas-Osorio, I.; Aburto-Oropeza, O. Hydroacoustics as a Tool to Examine the Effects of Marine Protected Areas and Habitat Type on Marine Fish Communities. *Sci. Rep.* **2018**, *8*, 47. [CrossRef]
89. Salvatat, J.; Bez, N.; Habasque, J.; Lebourges-Dhaussy, A.; Lopes, C.; Roudaut, G.; Simier, M.; Travassos, P.; Vargas, G.; Bertrand, A. Comprehensive Spatial Distribution of Tropical Fish Assemblages from Multifrequency Acoustics and Video Fulfils the Island Mass Effect Framework. *Sci. Rep.* **2022**, *12*, 8787. [CrossRef]
90. Villalobos-Ortíz, H.; Zwolinski, J.P.; Godínez-Pérez, C.A.; González-Máynez, V.E.; Mayorga-Martínez, M.; Michaels, W.L.; Palacios-Higuera, M.S.; Rubio-Rodríguez, U.; Sarmiento-Lezcano, A.N.; Demer, D.A. A Practical Approach to Monitoring Marine Protected Areas. *Oceanography* **2021**, *4*, 32–43. [CrossRef]
91. Jones, D.R.; Wilson, C.D.; de Robertis, A.; Rooper, C.N.; Weber, T.; Butler, J.L. Evaluation of Rockfish Abundance in Untrawlable Habitat: Combining Acoustic and Complementary Sampling Tools. *Fish. Bull.* **2012**, *110*, 332–343.
92. Xie, X.; Zhang, H.; Wang, C.; Wu, J.; Wei, Q.; Du, H.; Li, J.; Ye, H. Are River Protected Areas Sufficient for Fish Conservation? Implications from Large-Scale Hydroacoustic Surveys in the Middle Reach of the Yangtze River. *BMC Ecol.* **2019**, *19*, 42. [CrossRef] [PubMed]
93. Mayorga-Martínez, M.; Bello-Pineda, J.; Perales-Valdivia, H.; Pérez-España, H.; Heyman, W. Characterizing Geomorphology of Mesophotic Coral Reef Ecosystems in the Southwestern Gulf of Mexico: Implications for Conservation and Management. *Front. Mar. Sci.* **2021**, *8*, 639359. [CrossRef]
94. Cisneros-Montemayor, A.M.; Vincent, A. Science, Society, and Flagship Species: Social and Political History as Keys to Conservation Outcomes in the Gulf of California. *Ecol. Soc.* **2016**, *21*, 9. [CrossRef]
95. de Anda Montañez, J.A.; García-de-León, F.J.; Zenteno-Savín, T.; Balart-Paez, E.; Méndez-Rodríguez, L.C.; Bocanegra-Castillo, N.; Martínez-Aguilar, S.; Campos-Dávila, L.; Román-Rodríguez, M.J.; Valenzuela-Quiñones, F.; et al. *Estado de Salud y Estatus de Conservación de La(s) Población(Es) de Totoaba (Totoaba macdonaldi) En El Golfo de California: Una Especie En Peligro de Extinción*; Proyecto No. HK050; Centro de Investigaciones Biológicas del Noroeste, S.C. La Paz, Baja California Sur. Informe Final, SNIB-CONABIO: Tlalpan, México, 2013; p. 113.
96. Lynch, A.; Cooke, S.J.; Deines, A.M.; Bower, S.D.; Bunnell, D.B.; Cowx, I.G.; Nguyen, V.M.; Nohner, J.; Phouthavong, K.; Riley, B.; et al. The Social, Economic, and Environmental Importance of Inland Fish and Fisheries. *Environ. Rev.* **2016**, *24*, 115–121. [CrossRef]
97. Pollom, R.A.; Rose, G.A. A Global Review of the Spatial, Taxonomic, and Temporal Scope of Freshwater Fisheries Hydroacoustics Research. *Environ. Rev.* **2016**, *24*, 333–347. [CrossRef]

98. Linares, E.O.; Guillard, J.; Schneider, P.; Caballero-Caballero, P.I.; Gerlotto, F. Hydroacoustic Surveys as Contribution to the Study of Spawning Aggregations of Nassau Grouper (*Epinephelus striatus*) in the Yucatan. In Proceedings of the IEEE/OES Acoustics in Underwater Geosciences Symposium, Rio de Janeiro, Brazil, 24–26 July 2013; pp. 1–4. [CrossRef]
99. Rowell, T.J.; Demer, D.A.; Aburto-Oropeza, O.; Cota-Nieto, J.J.; Hyde, J.R.; Erisman, B.E. Estimating Fish Abundance at Spawning Aggregations from Courtship Sound Levels. *Sci. Rep.* **2017**, *7*, 3340. [CrossRef]
100. Lin, P.; Chen, L.; Gao, X.; Wang, C.; Gao, X.; Kang, M. Spatiotemporal Distribution and Species Composition of Fish Assemblages in the Transitional Zone of the Three Gorges Reservoir, China. *Water* **2020**, *12*, 3514. [CrossRef]
101. Simmonds, J.; MacLennan, D.N. *Fisheries Acoustics: Theory and Practice*, 2nd ed.; Blackwell Science Ltd.: New York, NY, USA, 2005; p. 472.
102. Love, R. Measurements of Fish Target Strength: A Review. *Fish. Bull.* **1971**, *69*, 703–715.
103. MacLennan, D.N. Acoustical Measurement of Fish Abundance. *J. Acoust. Soc. Am.* **1990**, *87*, 1–15. [CrossRef]
104. Ona, E. Physiological Factors Causing Natural Variations in Acoustic Target Strength of Fish. *J. Mar. Biol. Assoc. U. K.* **1990**, *70*, 107–127. [CrossRef]
105. Foote, K.G. Averaging of Fish Target Strength Functions. *J. Acoust. Soc. Am.* **1980**, *67*, 504–515. [CrossRef]
106. Liu, J.; Setiazi, H.; Yuk, P.-S. Fisheries Hydroacoustic Assessment: A Bibliometric Analysis and Direction for Future Research towards a Blue Economy. *Reg. Stud. Mar. Sci.* **2023**, *60*, 102838. [CrossRef]
107. ICES. Report of the Workshop on Survey Design and Data Analysis (WKSAD). In Proceedings of the Workshop on Survey Design and Analysis [WKSAD], Sète, France, 9–13 May 2005.
108. Trenkel, V.M.; Berger, L.; Bourguignon, S.; Doray, M.; Fablet, R.; Massé, J.; Mazauric, V.; Poncelet, C.; Quemener, G.; Scalabrin, C.; et al. Overview of Recent Progress in Fisheries Acoustics Made by Ifremer with Examples from the Bay of Biscay. *Aquat. Living Resour.* **2009**, *22*, 433–445. [CrossRef]
109. Harley, S.J.; Myers, R.A.; Dunn, A. Is Catch-per-Unit-Effort Proportional to Abundance? *Can. J. Fish. Aquat. Sci.* **2001**, *58*, 1760–1772. [CrossRef]
110. Martínez-Aguilar, S.; de Anda-Montañez, J.A.; Arreguín-Sánchez, F. Constant Harvest Rate for the Pacific Sardine (*Sardinops caeruleus*) Fishery in the Gulf of California Based on Catchability-at-Length Estimations. *Fish. Res.* **2009**, *99*, 74–82. [CrossRef]
111. Csirke, J. Small Schoalling Pelagic Fish Stocks. In *Fish Population Dynamics*, 2nd ed.; Gulland, J.A., Ed.; John Wiley: London, UK, 1988; pp. 271–302.
112. Arreguín-Sánchez, F. Catchability: A Key Parameter for Fish Stock Assessment. *Rev. Fish Biol. Fish.* **1996**, *6*, 221–242. [CrossRef]
113. Hilborn, R.; Walters, C.J. *Quantitative Fisheries Stock Assessment: Choice, Dynamics and Uncertainty*; Chapman and Hall: Boston, MA, USA, 1992; p. 470. [CrossRef]
114. Methot, R.D., Jr. Synthesis Model: An Adaptable Framework for Analysis of Diverse Stock Assessment Data. *Bull. Int. North Pac. Fish. Comm.* **1990**, *50*, 259–277.
115. Fournier, D.; Archibald, C.P. A General Theory for Analyzing Catch at Age Data. *Can. J. Fish. Aquat. Sci.* **1982**, *39*, 1195–1207. [CrossRef]
116. Polacheck, T.; Hilborn, R.; Punt, A. Fitting Surplus Production Models: Comparing Methods and Measuring Uncertainty. *Can. J. Fish. Aquat. Sci.* **1993**, *50*, 2597–2607. [CrossRef]
117. MacLennan, D.N.; Fernandes, P.G.; Dalen, J. A Consistent Approach to Definitions and Symbols in Fisheries Acoustics. *ICES J. Mar. Sci.* **2002**, *59*, 365–369. [CrossRef]
118. Haddon, M. *Modelling and Quantitative Methods in Fisheries*, 2nd ed.; Chapman and Hall: Boston, MA, USA; CRC: Boca Raton, FL, USA, 2011.
119. Petitgas, P. Geostatistics in Fisheries Survey Design and Stock Assessment: Models, Variances and Applications. *Fish Fish.* **2001**, *2*, 231–249. [CrossRef]
120. Mello, L.G.S.; Rose, G.A. Using Geostatistics to Quantify Seasonal Distribution and Aggregation Patterns of Fishes: An Example of Atlantic Cod (*Gadus morhua*). *Can. J. Fish. Aquat. Sci.* **2005**, *62*, 659–670. [CrossRef]
121. Pennington, M. Some Statistical Techniques for Estimating Abundance Indices from Trawl Surveys. *Fish. Bull.* **1986**, *84*, 519–525.
122. Syrjala, S.E. Critique on the Use of the Delta Distribution for the Analysis of Trawl Survey Data. *ICES J. Mar. Sci.* **2020**, *57*, 831–842. [CrossRef]
123. Jolly, G.; Hampton, I. A Stratifid Random Transect Design for Acoustic Surveys of Fish Stock. *Can. J. Fish. Aquat. Sci.* **1990**, *47*, 1282–1291. [CrossRef]
124. Lo, N. Biomass and Reproduction of Pacific Sardine (*Sardinops sagax*) off the Pacific Northwestern United States, 2003–2005. *Fish. Bull.* **2010**, *108*, 174–192.
125. Lo, N.G.-H.; Hunter, J.R.; Moser, H.G.; Smith, P.E. A Daily Fecundity Reduction Method of Biomass Estimation with Application to Dover Sole *Microstomus pacificus*. *Bull. Mar. Sci.* **1993**, *53*, 842–863.
126. Lo, N.G.-H.; Green-Ruiz, Y.A.; Cervantes, M.J. Egg Production and Spawning Biomass of Pacific Sardine (*Sardinops sagax*) in 1994, Determined by the Daily Egg Production Method. *Calif. Coop. Ocean Fish. Investig. Rep.* **1996**, *37*, 160–174.
127. Jacobson, L.; Konno, E.S.; Pertierra, J.P. Status of Pacific Mackerel and Trends in Biomass, 1978–1993. *Calif. Coop. Ocean Fish. Investig. Rep.* **1994**, *35*, 36–39.
128. Conser, R.J.; Kevin, T.H.; Crone, P.R.; Lo, N.G.-H.; Berger, D. *Stock Assessment of Pacific Sardine with Management Recommendations for 2003 Executive Summary*; Pacific Fishery Management Council: Portland, OR, USA, 2002.

129. Jacobson, L.; Lo, N.G.-H.; Barnes, J. A Biomass-Based Assessment Model for Northern Anchovy, *Engraulis mordax*. *Fish. Bull.* **1994**, *92*, 711–724.
130. Green-Ruiz, Y.A.; Cotero-Altamirano, C.E. Spawning Biomass of the Northern Anchovy (Engraulis Mordax) in the Gulf of California during 1992. *Cienc. Pesq.* **2009**, *17*, 27–96.
131. Nevárez-Martínez, M.O.; Morales-Bojórquez, E.; De Los Angeles Martínez-Zavala, M.; Villalobos-Ortíz, H.; Luquin-Covarrubias, M.; González-Máynez, V.E.; López-Martínez, J.; Santos-Molina, J.P.; Ornelas-Vargas, A.; Delgado-Vnces, F. An Integrated Catch-at-Age Model for Analyzing the Variability in Biomass of Pacific Sardine (*Sardinops sagax*) from the Gulf of California, Mexico. *Front. Mar. Sci.* **2023**, *10*, 940083. [CrossRef]
132. Lawson, G.L.; Rose, G.A. The Importance of Detectability to Acoustic Surveys of Semi-Demersal Fish. *ICES J. Mar. Sci.* **1999**, *56*, 370–380. [CrossRef]
133. Legault, C.; Restrepo, V.R. A Flexible Forward Age-Structured Assessment Program. *ICCAT. Col. Vol. Sci. Pap.* **1998**, *49*, 246–253.
134. Bull, B.; Francis, R.; Dunn, A. *CASAL (C++ Algorithmic Stock Assessment Laboratory) CASAL User Manual v2.30-2012/03/21*; NIWA Technical Report; NIWA: Auckland, New Zealand, 2012; p. 135.
135. Methot, R.D., Jr.; Wetzel, C.R. Stock Synthesis: A Biological and Statistical Framework for Fish Stock Assessment and Fishery Management. *Fish. Res.* **2013**, *142*, 86–99. [CrossRef]
136. Anon. Assessment Model for Alaska Description of GUI and Instructions. 2015. Available online: https://github.com/NMFS-toolbox/AMAK/blob/master/docs/AMAK%20Documentation.pdf (accessed on 4 September 2024).
137. Cope, J. Implementing a Statistical Catch-at-Age Model (Stock Synthesis) as a Tool for Deriving Overfishing Limits in Data-Limited Situations. *Fish. Res.* **2013**, *142*, 3–14. [CrossRef]
138. Cope, J.; MacCall, A.; Monk, M. *Data-Moderate Stock Assessments for Brown, China, Copper, Sharpchin, Stripetail, and Yellowtail Rockfishes and English and Rex Soles in 2013*; Pacific Fishery Management Council: Portland, OR, USA, 2015; Volume 97220, p. 298.
139. Wetzel, C.R.; Punt, A.E. Performance of a Fisheries Catch-at-Age Model (Stock Synthesis) in Datalimited Situations. *Mar. Freshw. Res.* **2011**, *62*, 927–936. [CrossRef]
140. Stock, B.C.; Miller, T.J. The Woods Hole Assessment Model (WHAM): A General State-Space Assessment Framework That Incorporates Time- and Age-Varying Processes via Random Effects and Links to Environmental Covariates. *Fish. Res.* **2021**, *240*, 105967. [CrossRef]
141. Fournier, D.; Hampton, J.; Sibert, J.R. MULTIFAN-CL: A Length-Based, Age-Structured Model for Fisheries Stock Assessment, with Application to South Pacific Albacore, Thunnus Alalunga. *Can. J. Fish. Aquat. Sci.* **1998**, *55*, 2105–2116. [CrossRef]
142. Benoit-Bird, K.J.; Welch, T.P.; Waluk, C.M.; Barth, J.A.; Wangen, I.; McGill, P.; Okuda, C.; Hollinger, G.A.; Sato, M.; McCammon, S. Equipping an Underwater Glider with a New Echosounder to Explore Ocean Ecosystems. *Limnol. Oceanogr. Methods* **2018**, *16*, 734–749. [CrossRef]
143. Goulon, C.; Le Meaux, O.; Vincent-Falquet, R.; Guillard, J. Hydroacoustic Autonomous Boat for Remote Fish Detection in LakE (HARLE), an Unmanned Autonomous Surface Vehicle to Monitor Fish Populations in Lakes. *Limnol. Oceanogr. Methods* **2021**, *19*, 280–292. [CrossRef]
144. Eriksen, E.; Gjøsæter, H.; Prozorkevich, D.; Shamray, E.; Dolgov, A.; Skern-Mauritzen, M.; Stiansen, J.E.; Kovalev, Y.; Sunnana, K. From Single Species Surveys towards Monitoring of the Barents Sea Ecosystem. *Prog. Oceanogr.* **2018**, *166*, 4–14. [CrossRef]
145. Izquierdo-Peña, V.; Lluch-Cota, S.E.; Chavez, F.P.; Lluch-Belda, D.; Morales-Bojórquez, E.; Ponce-Díaz, G. Is There a Future in the Sustainability Certification of Sardine and Anchovy Fisheries? *Fisheries* **2020**, *45*, 554–560. [CrossRef]
146. Diario Oficial de La Federación. *Plan de Manejo Pesquero Para La Pesquería de Pelágicos Menores (Sardina, Anchovetas, Macarela y Afines) Del Noroeste de México*; Diario Oficial de La Federación: Cuauhtémoc, Mexico, 2012.
147. Diario Oficial de La Federación. *NORMA Oficial Mexicana NOM-003-SAG/PESC-2018*; Diario Oficial de La Federación: Cuauhtémoc, México, 2019.
148. Chu, D.; Parker-Stetter, S.; Hufnagle, L., Jr.; Thomas, R.; Getsiv-Clemons, J.; Gauthier, S.; Stanley, C. 2018 Unmanned Surface Vehicle (Saildrone) Acoustic Survey off the West Coasts of the United States and Canada. In Proceedings of the OCEANS 2019 MTS/IEEE, Seattle, WA, USA, 27–31 October 2019. [CrossRef]
149. de Robertis, A.; Lawrence-Slavas, N.; Jenkins, R.; Wangen, I.; Mordy, C.W.; Meinig, C.; Levine, M.; Peacock, D.; Tabisola, H. Long-Term Measurements of Fish Backscatter from Saildrone Unmanned Surface Vehicles and Comparison with Observations from a Noise-Reduced Research Vessel. *ICES J. Mar. Sci.* **2019**, *76*, 2459–2470. [CrossRef]

fishes

MDPI

Article

Acoustic Target Strength Measurement of *Larmichthys crocea* Based on the Kirchhoff-Ray Mode Model

Junliang Meng [1,2], Yong Tang [2], Lizhi Sun [1], Longshan Lin [1], Yuan Li [1], Xing Miao [1], Shigang Liu [1,*] and Puqing Song [1,*]

[1] Third Institute of Oceanography, Ministry of Natural Resources, Xiamen 361005, China; mjunliangss@163.com (J.M.)

[2] College of Marine Living Resource Sciences and Management, Shanghai Ocean University, Shanghai 201306, China

* Correspondence: liushigang@tio.org.cn (S.L.); songpuqing@tio.org.cn (P.S.)

Abstract: *Larmichthys crocea* (*L. crocea*) is an economically important fish species mainly distributed off the coast of China. In this study, 11 *L. crocea* samples of different body lengths were collected from aquaculture cages in the East China Sea to measure the acoustic target strength (TS). Using the Kirchhoff-Ray mode (KRM) model, the directional TS of *L. crocea* was measured at frequencies of 70 kHz, 120 kHz, and 200 kHz. Furthermore, the relationships between TS and body length were determined using the least squares method and the standard b_{20} equation. The results showed that the TS of *L. crocea* varied with tilt angle, frequency, and body length. For tilt angles ranging from (−5°, 15°), the equation for fitting TS and body length using the least squares method were $TS = 32.99 \cdot \log_{10}L - 87.36$ (70 kHz), $TS = 33.26 \cdot \log_{10}L - 87.77$ (120 kHz), and $TS = 39.46 \cdot \log_{10}L - 95.51$ (200 kHz). They were expressed in the standard b_{20} equation as $TS = 20 \cdot \log_{10}L - 71.16$ (70 kHz), $TS = 20 \cdot \log_{10}L - 71.23$ (120 kHz), and $TS = 20 \cdot \log_{10}L - 71.24$ (200 kHz). For tilt angles ranging from (0°, 10°), the equation for fitting TS and body length using the least squares method is $TS = 28.69 \cdot \log_{10}L - 81.71$ (70 kHz), $TS = 32.30 \cdot \log_{10}L - 86.44$ (120 kHz), and $TS = 45.87 \cdot \log_{10}L - 103.73$ (200 kHz). They were expressed in the standard b_{20} equation as $TS = 20 \cdot \log_{10}L - 70.88$ (70 kHz), $TS = 20 \cdot \log_{10}L - 71.10$ (120 kHz), and $TS = 20 \cdot \log_{10}L - 71.48$ (200 kHz). Moreover, the 18–300 kHz spectral curve showed a decreasing trend in the frequency range from 18 kHz to 30 kHz, while maintaining relatively stable fluctuations in the other frequency ranges. These findings offer a comprehensive understanding of the scattering characteristics of *L. crocea* and provide a reliable reference of TS for the fishery acoustic assessment of *L. crocea*.

Keywords: acoustic; target strength; *Larmichthys crocea*; kirchhoff-ray mode; tilt angle

Key Contribution: The KRM model was used to analyze the variation patterns of *L. crocea* TS under different frequencies, tilt angles, and body length conditions. Meanwhile, the relationship equations between the dorsal TS and body length of *L. crocea* at 70 kHz, 120 kHz, and 200 kHz were obtained.

1. Introduction

Larmichthys crocea is a warm-temperate migratory fish species that predominantly inhabits the lower and middle layers of the sea at depths of up to 60 m along coastal areas [1]. Its distribution spans the northwestern Pacific Ocean, from central Vietnam to South Korea and Japan [2]. The highest concentrations of *L. crocea* are found along the coast of China, particularly in the Yellow Sea and East China Sea [3]. However, the abundance of *L. crocea* sharply declined after the 1980s in China due to overfishing [4]. In response, China has implemented numerous conservation measures to protect and rehabilitate wild *L. crocea* populations, such as establishing protected areas and stock enhancement and release [5]. Therefore, accurate and timely assessment of *L. croccea* stock is essential for developing effective management strategies [6].

Traditional fishery resource assessment surveys often rely on direct fishing methods, which are susceptible to biases from fishing gear selectivity, fishing efficiency, and sampling area size, resulting in lower assessment efficiency [7]. In contrast, fishery acoustic offers several advantages including rapid, continuous, non-invasive, and low-cost data collection. This technique is widely applicable in oceans [8,9], rivers [10–12], lakes [13], and reservoirs [14,15] and has garnered significant global attention. In fishery acoustic assessments, an essential physical parameter used to measure a fish's reflective capability is target strength (TS) [16]. TS describes the acoustic properties of a target based on echo intensity and is used to convert echo integration values into key parameters to determine fish size and abundance [17].

Due to the diverse external shapes of different fish species, even fish with similar shapes may have significantly different internal structures. As a result, fish bodies are considered complex acoustic scattering targets, posing unique challenges for acoustic analysis [18,19]. Broadly, the factors affecting TS can be categorized into internal and external factors. The internal factors are mainly differences in fish body structure, including fish size, the acoustic scattering properties of tissues and organs, the presence of swim bladders, and differences in swim bladder structure [20]. External factors include the frequency of the acoustic waves used for detection, the depth at which the fish are located, and the relative posture and tilt angles between the transducer and the fish body [21,22]. These factors collectively impact how fish scatter acoustic signals, influencing the measurement and interpretation of TS in acoustic assessments of fish populations. In coastal fishery resource surveys, echosounders commonly operate at frequencies of 70 kHz, 120 kHz, and 200 kHz, which are specifically chosen for their ability to generate wavelengths shorter than the typical lengths of the fish being detected [23]. This configuration enhances the resolution and precision of sonar-based data acquisition, which is crucial for informing sustainable management strategies [24].

Currently, techniques for measuring TS can be broadly categorized into the controlled experimental method [22], in situ measurement method [25], and theoretical modeling method. The theoretical modeling method includes the sphere model [26], prolate spheroidal model [27], deformed cylinder model [28], Kirchhoff model [29], distorted wave Born approximation model [30], and finite element method [31]. Theoretical modeling is flexible and convenient, capable of simulating TS values under various conditions without being limited by practical constraints such as site facilities, making it particularly suitable for estimating the TS of fish species that cannot be measured directly in situ [19]. Moreover, with increasing model certainty, theoretical modeling has become one of the mainstream methods in TS research [32]. The KRM model, in particular, provides a good approximation for fish bodies and swim bladders, accurately reflecting the situation of TS [33,34].

Foot [35] and Clay [29] proposed models based on the Kirchhoff approximation, which better approximates the fish body and swim bladder. The KRM model is suitable for modeling at majority frequencies [36]; it works by segmenting the fish body and swim bladder into several small elements, calculating the TS for each element, and then summing them to obtain the overall TS. While the Kirchhoff model can accurately reflect TS [37], it requires careful measurements of the fish body and swim bladder shapes. Currently, TS for various fish species was measured using the KRM model, including yellowfin tuna (*Thunnus albacares*) [38], skipjack tuna (*Katsuwonus pelamis*) [38], blackhead seabream (*Acanthopagrus schlegelii*) [39], chub mackerel (*Scomber japonicus*) [40], silver carp (*Hypophthalmichthys molitrix*), and bighead carp (*Aristichthys nobilis*) [41].

Despite the *L. crocea* being one of the most economically important fish species in China's coastal waters, there is little research on its TS. Luan et al. [42] and Shang [43] using a controlled experimental method, conducted preliminary research on the TS of *L. crocea*, but these studies cannot be directly applied to the acoustic assessment of nearshore *L. crocea* resources due to inappropriate transducer angles or frequencies. Therefore, accurate TS measurements are essential for conducting effective fishery acoustic assessments of *L. crocea*.

In this study, the KRM model was used to analyze the variation patterns of *L. crocea* TS under different frequencies, tilt angles, and body length. The equations for the relationship between body length and TS were analyzed by measuring the TS of *L. crocea* at different body lengths. Concurrently, the spectral curves of *L. crocea* to TS were analyzed at 18–300 kHz. The results of this study provide a comprehensive understanding of the scattering characteristics of *L. crocea* and also provide a reliable reference of TS for fishery acoustic assessment of *L. crocea*.

2. Materials and Methods

2.1. Fish Sample Collection

A total of 11 *L. crocea* samples, varying in body length, were collected from deep-water aquaculture nets in the East China Sea. All samples were obtained in site, and ice cooling was used to ensure the samples were kept fresh. After the samples were brought back to the laboratory, biological parameters were measured, including body length and weight.

2.2. Obtaining X-Ray Images of *L. crocea*

X-ray images of each of the 11 *L. crocea* samples were captured using a digital medical diagnostic X-radiography system (TITAN2000, 49 Kv, 25 mA, 16 mAs) within 12 h of collection. After obtaining the X-ray images, the morphological parameters of each sample were measured using Adobe Photoshop (2024). These parameters included the total length, width, and height of the fish body and swim bladder, as well as the tilt angle, which is the major axis angle between the swim bladder and the fish. In accordance with the KRM model requirements, the X-ray images were segmented into approximately equidistant slices, and the coordinates of each segment were measured (Figure 1).

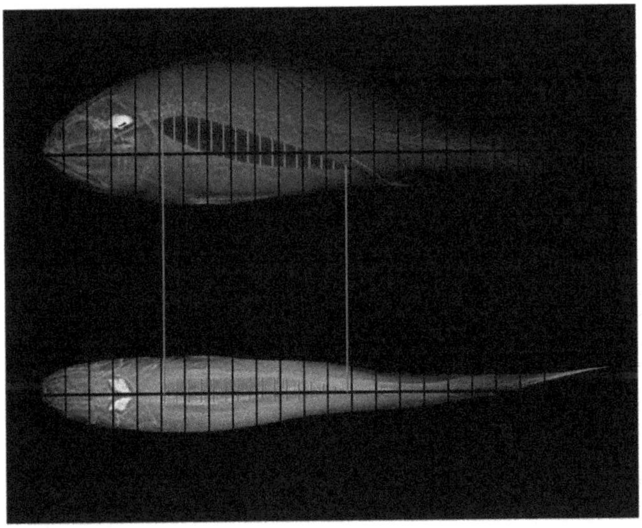

Figure 1. Morphological parameters measurement of *Larmichthys crocea*. **Top panel**: lateral view, **bottom panel**: dorsal view. The black line is the line segment dividing the fish body into approximately equidistant slices, and the blue line is the line segment dividing the swim bladder into approximately equidistant slices.

2.3. Kirchhoff-Ray Mode Model

The TS of *L. crocea* was measured using the KRM model. The specific method of establishing the KRM model consists of the following three steps. First, the fish body and swim bladder are sliced into approximate equidistant vertical segments, which are decomposed into several approximate cylindrical bodies. Then, the scattering strength of each cylindrical body of the fish body and swim bladder is calculated separately. Finally, the scattering strengths of the fish body and swim bladder are integrated to obtain the total TS of the target fish. The geometric construction of the KRM model is shown in Figure 2 [44].

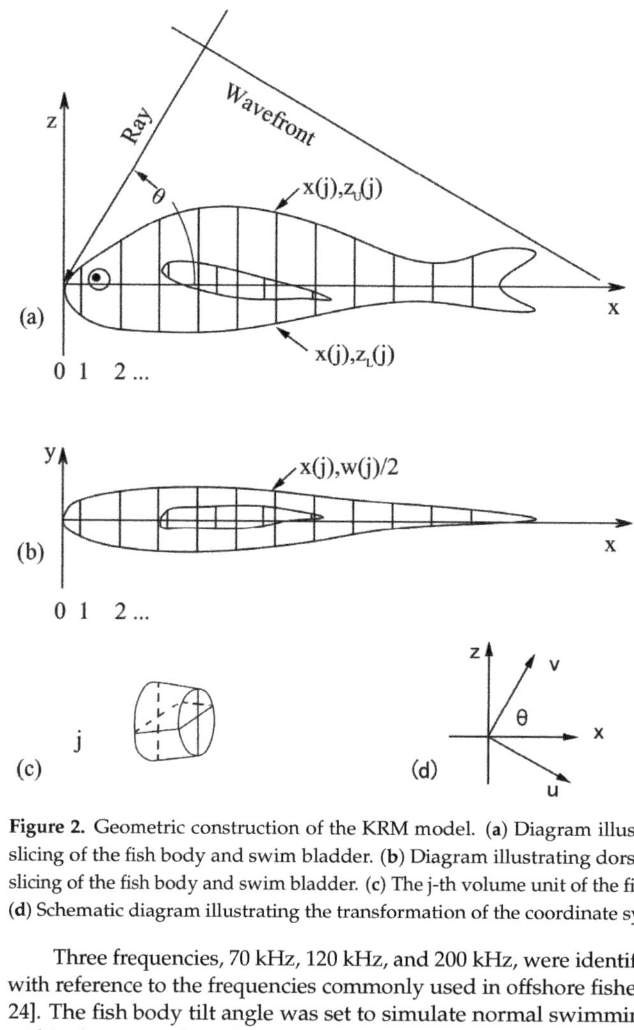

Figure 2. Geometric construction of the KRM model. (**a**) Diagram illustrating lateral equidistant slicing of the fish body and swim bladder. (**b**) Diagram illustrating dorsal and ventral equidistant slicing of the fish body and swim bladder. (**c**) The j-th volume unit of the fish body and swim bladder. (**d**) Schematic diagram illustrating the transformation of the coordinate system from (x, z) to (u, v).

Three frequencies, 70 kHz, 120 kHz, and 200 kHz, were identified for TS measurement with reference to the frequencies commonly used in offshore fishery resource surveys [23, 24]. The fish body tilt angle was set to simulate normal swimming conditions and is set within the range of $-50°$ to $50°$ ($0°$ when the sound wave is perpendicular to the fish body, with negative angles indicating downward tilting of the fish head) [45]. The tilt angle ranges of the fish body were selected based on monitoring the daily activities of the fish, resulting in truncated normal tilt angle ranges: ($-5°$, $15°$) and ($0°$, $10°$) (mean, standard deviation) [46]. These two classic tilt angle ranges were used to calculate the average TS of *L. crocea*. The seawater environmental parameters and typical acoustic parameters of marine fish required for the KRM model TS calculations are shown in Table 1 [44]. For detailed calculations of the KRM model, refer to Clay [44]. The KRM model was built using MATLAB software (R2023b).

Table 1. Seawater environmental parameters and typical marine fish of acoustic parameters. Density contrast and sound speed contrast are both relative to sea water.

Medium	Density (kg/m^3)	Sound Speed (m/s)	Density Contrast	Sound Speed Contrast
Sea water	1030	1520	-	-
Fish body	1070	1570	1.04	1.03
Swim bladder	1.24	345	0.001	0.23

2.4. Calculation and Evaluation of TS

The equation for calculating the TS of a fish [44] is

$$TS = 10 \cdot \log_{10}(L_s + L_b)^2 \quad (1)$$

Here, L_s represents the total scattering length of the swim bladder and L_b represents the total scattering length of the fish body.

The regression equation describing the relationship between TS and body length using the least squares method is

$$TS = a \cdot \log_{10} L + b \quad (2)$$

where L represents the body length of the fish (cm), a is the slope of the regression equation, and b denotes the intercept. The standard b_{20} equation [47] is

$$TS = 20 \cdot \log_{10} L + b_{20} \quad (3)$$

where b_{20} represents the intercept and a = 20.

The average TS was calculated over two classic tilt angle ranges, (−5°, 15°) and (0°, 10°), with 70 kHz, 120 kHz, and 200 kHz, respectively.

To obtain the continuous variation in *L. crocea* TS with frequency, spectral curves were plotted. The frequency range was modeled from 18 to 300 kHz at 1 kHz increments.

Two methods of the least squares method equation and standard b_{20} equation were used to fit the relationship between TS and body length. TS at different frequencies under tilt angle ranges of (−5°, 15°) and (0°, 10°), along with the corresponding body lengths, were fitted using the least squares method and the standard b_{20} equation.

3. Results and Analysis

3.1. Biological Sampling and Measurement

The biological parameters, including body length, body weight, and tilt angle between the swim bladder and body of the 11 *L. crocea* samples, are shown in Table 2. The body length of *L. crocea* ranges from 13.0 cm to 23.6 cm, with a mean of 17.95 cm, and the body weight ranges from 32.1 g to 238.4 g, with a mean of 106.33 g. The tilt angle between the swim bladder and the fish body was measured using X-ray images and ranges from −12° to −17°, with a mean of −14.6°.

Table 2. Biological parameters of 11 *Larmichthys crocea* samples. Tilt angle is the major axis angle between the swim bladder and the fish.

Serial Number	Body Length (cm)	Body Weight (g)	Tilt Angle (°)
1	13.0	32.1	−16
2	13.8	38.6	−17
3	14.9	52.4	−14
4	15.5	59.2	−15
5	17.0	81.6	−14
6	17.9	104.3	−14
7	18.9	100.6	−16

Table 2. *Cont.*

Serial Number	Body Length (cm)	Body Weight (g)	Tilt Angle (°)
8	20.0	150.2	−15
9	21.1	138.4	−14
10	21.7	173.8	−12
11	23.6	238.4	−14

3.2. Variation in L. crocea TS with the Tilt Angle at Different Frequencies

The TS of 11 *L. crocea* samples with tilt angles at different frequencies were measured using the KRM model. The results show that TS varies with tilt angle for *L. crocea* of different sizes (Figure 3, samples No. 1, 6, and 11 are shown). However, the TS of the swim bladder (TS_s) closely overlaps with the total TS of the fish (TS_t), indicating that the swim bladder is the main source of acoustic scattering in *L. crocea* and plays a decisive role in determining TS.

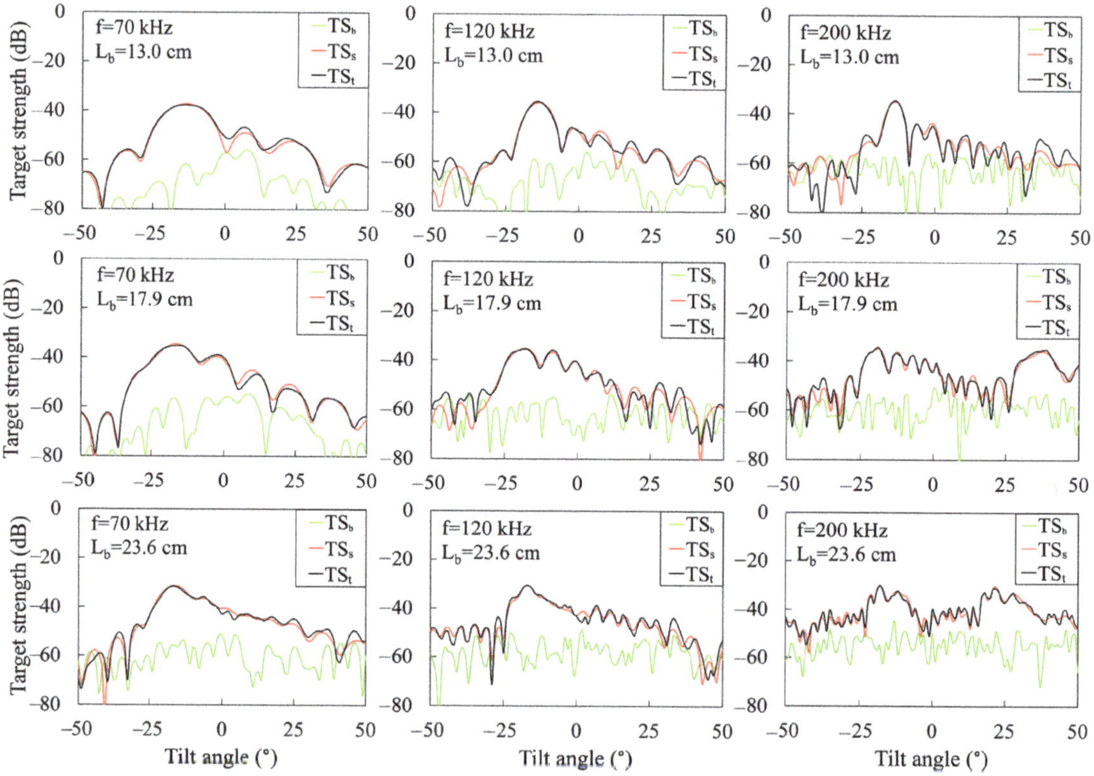

Figure 3. Variation in the TS of *Larmichthys crocea* with a tilt angle at different frequencies. L_b represents body length, TS_t represents total TS, TS_s represent swim bladder TS, and TS_b represents fish body TS.

At frequencies of 70 kHz, 120 kHz, and 200 kHz, the TS of the fish exhibits a multi-peaked distribution with respect to the tilt angle. As the frequency increases, the sensitivity of the TS to tilt angle variation increases, which leads to an increase in the number of peaks.

In addition, the tilt angle corresponding to the maximum TS increases with the frequency. At 70 kHz, the maximum TS occurs between tilt angles of −18° and −12°, with an average TS of 15.2°. At 120 kHz, the maximum TS primarily occurs between −19° and

−12°, and the average TS is 15.8°. At 200 kHz, the maximum TS occurs between −20° and −13°, and the average TS is 15.9°. These findings suggest that the effect of fish posture, which corresponds to fish behavior, on TS becomes more pronounced at higher frequencies.

3.3. Relationship Between L. crocea TS and Body Length Across Different Frequencies

For tilt angle ranges (−5°, 15°), at 70 kHz, the average TS is −46.23 dB, with an average b_{20} of −71.16 dB. At 120 kHz, the average TS is −46.30 dB, with an average b_{20} of −71.23 dB. At 200 kHz, the average TS is −46.24 dB, with an average b_{20} of −71.24 dB. For tilt angle ranges (0°, 10°), at 70 kHz, the average TS is −45.95 dB, with an average b_{20} of −70.88 dB. At 120 kHz, the average TS is −46.17 dB, with an average b_{20} of −71.10 dB. At 200 kHz, the average TS is −46.55 dB, with an average b_{20} of −71.48 dB (Tables 3 and 4).

Table 3. TS of *Larmichthys crocea* at different frequencies and tilt angle ranges.

Serial Number	Mean TS at Tilt Angle Ranges of (−5°, 15°)			Mean TS at Tilt Angle Ranges of (0°, 10°)		
	Frequency/kHz					
	70 kHz	120 kHz	200 kHz	70 kHz	120 kHz	200 kHz
1	−49.62	−50.98	−51.60	−48.99	−50.32	−52.51
2	−50.83	−48.58	−51.28	−49.70	−48.59	−51.28
3	−49.94	−50.04	−50.26	−47.05	−51.04	−51.58
4	−46.87	−48.05	−47.38	−47.04	−46.87	−48.09
5	−46.12	−46.52	−45.99	−46.58	−47.10	−46.51
6	−45.85	−46.44	−46.34	−47.78	−45.97	−46.88
7	−47.18	−45.65	−45.02	−48.16	−45.33	−45.68
8	−43.29	−44.27	−42.75	−43.23	−43.37	−43.50
9	−43.85	−43.58	−42.21	−42.43	−44.41	−42.37
10	−42.37	−42.98	−42.92	−41.25	−41.76	−41.91
11	−42.59	−42.26	−42.92	−43.21	−43.09	−41.70

Table 4. The b_{20} of *Larmichthys crocea* at different frequencies and tilt angle ranges.

Serial Number	The b_{20} at Tilt Angle Ranges of (−5°, 15°)			The b_{20} at Tilt Angle Ranges of (0°, 10°)		
	70 kHz	120 kHz	200 kHz	70 kHz	120 kHz	200 kHz
1	−71.90	−73.26	−73.88	−71.26	−72.60	−74.79
2	−73.63	−71.38	−74.08	−72.50	−71.39	−74.08
3	−73.41	−73.50	−73.72	−70.51	−74.50	−75.04
4	−70.68	−71.86	−71.19	−70.84	−70.68	−71.89
5	−70.73	−71.13	−70.60	−71.19	−71.71	−71.12
6	−70.91	−71.49	−71.40	−72.84	−71.03	−71.94
7	−72.71	−71.18	−70.55	−73.69	−70.86	−71.21
8	−69.31	−70.29	−69.66	−69.25	−69.39	−69.52
9	−70.34	−70.06	−69.23	−68.91	−70.90	−68.85
10	−69.09	−69.71	−68.94	−67.98	−68.49	−68.63
11	−70.04	−69.72	−70.38	−70.67	−70.54	−69.16

The relationships between TS at different frequencies and corresponding body lengths with two tilt angle ranges (−5°, 15°) and (0°, 10°) are depicted in Figure 4. At all frequencies, TS showed an upward trend with increasing body length.

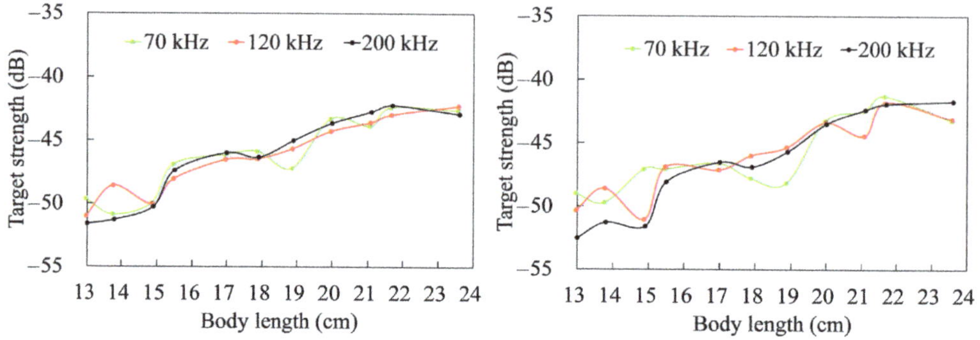

Figure 4. Variations in the TS of *Larmichthys crocea* with body length at different frequencies. The figure on the left shows the tilt angle (−5°, 15°). The figure on the right shows the tilt angle (0°, 10°).

Statistical analysis revealed the variations in TS with body length in *L. crocea* at different frequencies (Figures 5–7).

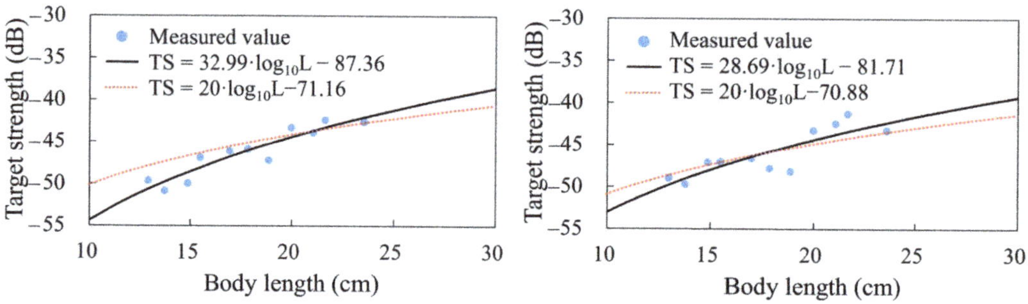

Figure 5. Fitting curve of the TS of *Larmichthys crocea* with body length at 70 kHz. The figure on the left shows the tilt angle (−5°, 15°). The figure on the right shows the tilt angle (0°, 10°).

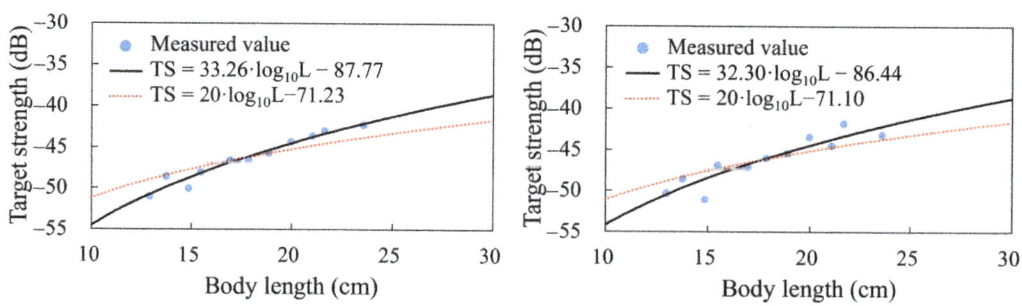

Figure 6. Fitting curve of the TS of *Larmichthys crocea* with body length at 120 kHz. The figure on the left shows the tilt angle (−5°, 15°). The figure on the right shows the tilt angle (0°, 10°).

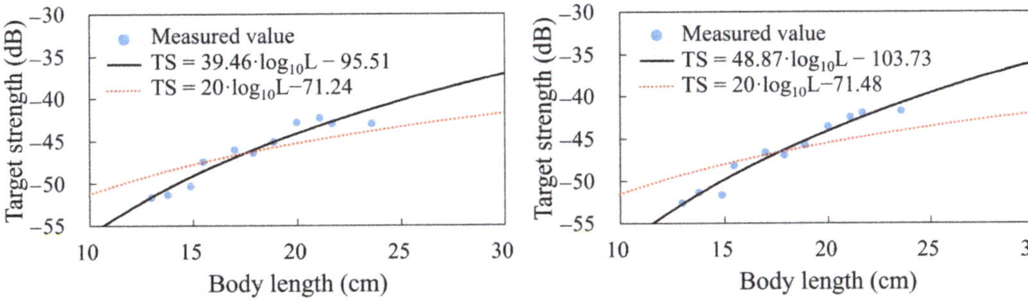

Figure 7. Fitting curve of the TS of *Larmichthys crocea* with body length at 200 kHz. The figure on the left shows the tilt angle (−5°, 15°). The figure on the right shows the tilt angle (0°, 10°).

3.4. Frequency Response of *L. crocea* to TS

To analyze the variation in *L. crocea* TS with frequency under broadband conditions, a sample fish (Number 6), which was closest to the average weight and body length of all 11 samples, was used for analysis (Table 1). Spectral curves of the TS between 18–300 kHz were obtained using the KRM model for tilt angle ranges of (−5°, 15°) and (0°, 10°) (Figure 8). Overall, the spectral curves show a similar trend of variation for both sets of tilt angles. The TS shows a decreasing trend in the frequency range of 18 kHz to 30 kHz, while it remains relatively stable at other frequencies.

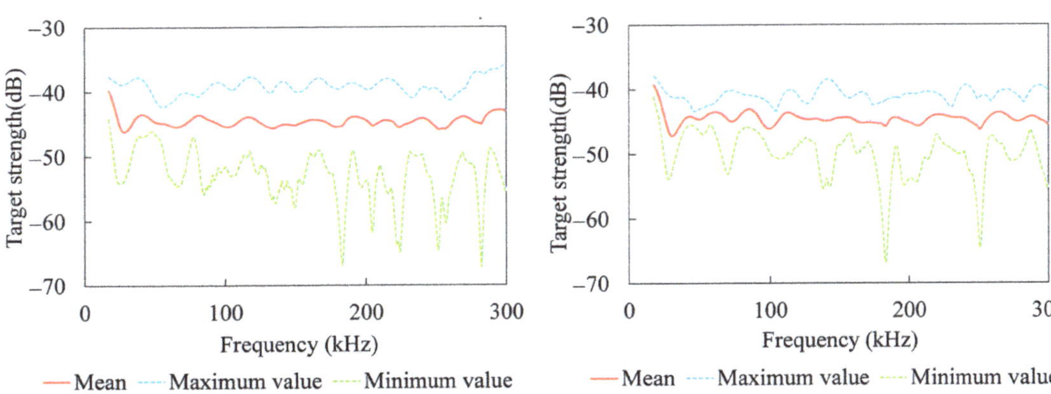

Figure 8. Frequency response of *Larmichthys crocea* to TS. The figure on the left was plotted with the tilt angle (−5°, 15°), and the figure on the right was plotted with the tilt angle (0°, 10°).

4. Discussion

In this study, the KRM model was used to measure the TS of *L. crocea* at different body lengths, tilt angles, and acoustic detection frequencies. Among these factors, changes in tilt angle have a particularly significant effect on TS compared to body length and frequency. For example, Sample No. 6 at 70 kHz has a maximum TS of −35.14 dB at a −16° tilt angle and a minimum of −78.84 dB at a −45° tilt angle (Figure 4). This study utilized two commonly used tilt angle ranges (−5°, 15°) and (0°, 10°), when calculating the average TS. Furusawa [48] proposed these tilt angle ranges and demonstrated that the average TS calculated in these ranges is in good agreement with the actual TS. The results of the study revealed minimal differences in the average TS obtained from the two tilt angle ranges at the same frequencies, with an average difference of 0.82 dB for 11 samples (Table 3). The differences in b_{20} values in the two tilt angle ranges are 0.28 dB (70 kHz), 0.13 dB (120 kHz), and 0.24 dB (200 kHz). These findings underscore the significant role of tilt angle variation on fish TS. Therefore, it is essential to conduct further research on how TS varies with tilt angle during different fish behaviors to improve the accuracy of TS measurements.

As the detection frequency increased, the number of secondary peaks in the TS increased (Figure 4). At three different frequencies, the average tilt angles where the TS reaches its maximum values are $-15.2°$, $-15.8°$, and $-15.9°$ (Figure 4), which are close to the average tilt angle of $-14.3°$ between the swim bladder and the fish body (Table 2). Similar results have been reported in other studies involving fish with swim bladders, confirming the significant impact of the swim bladder on TS [19,49,50]. Because the swim bladder is not a standard ellipsoid, there are differences in the tilt angles between the swim bladder and the fish body, as well as in the tilt angles where TS reaches its maximum values, which vary with detection frequency [51]. *L. crocea* are classified as swim bladder fish, with the swim bladder serving as the primary organ for acoustic scattering, contributing approximately 90% to 95% of the fish's scattering capability [18]. In contrast, other tissues and organs(such as bones, muscles, and viscera) make minimal contributions to the overall TS due to their density being similar to that of the surrounding water [52]. Therefore, considering the decisive role of the swim bladder in determining the TS of swim bladder fish species, factors such as the position, shape, and size of the swim bladder within the abdominal cavity, as well as influencing factors such as gastric fullness and gonadal development, are crucial in influencing fish TS [53]. For the same swim bladder fish species, even if they have similar body shapes, their TS can vary due to factors such as distribution in marine areas, temperature and salinity conditions, individual growth status, and body condition [54].

Body length is a primary factor affecting the TS of fish [6]. Therefore, studying the relationship between TS and body length in *L. crocea* is an important basis for acoustic assessment in fisheries. Luan et al. [42] using a controlled experimental method, measured the TS of *L. crocea* from different directions at frequencies of 50·kHz and 200 kHz, only the lateral TS equations were given: $TS = 24.8 \cdot \log_{10}L - 73.9$ (at 50 kHz) and $TS = 23.9 \cdot \log_{10}L - 71.3$ (200 kHz). However, typical fishery resource surveys are conducted from the dorsal direction of the fish. Shang [43] analyzed variations in the dorsal TS of *L. crocea* at 430 kHz with respect to tilt angle, body length, and weight using the controlled experimental method and an ellipsoid model. He derived equations for dorsal TS based on body length and weight as $TS = 23.5 \cdot \log_{10} L - 75.8$ and $TS = 8.8 \cdot \log_{10} G–64.4$. However, the frequencies commonly used in fisheries acoustic surveys are 38 kHz, 70 kHz, 120 kHz, and 200 kHz. The frequency 430 kHz is not commonly used in fisheries acoustic surveys, so these results cannot be used in the fishery acoustic assessment of nearshore *L. crocea* resources.

In this study, equations between dorsal TS and body length of *L. crocea* were obtained for three commonly used frequencies (70 kHz, 120 kHz, and 200 kHz) using the KRM model. To find the most suitable equation, we compared the least squares method with the standard b_{20} expression. The results indicate that both fitted equations exhibited some deviations at different acoustic frequencies and tilt angle ranges. Notably, the R^2 of the equation using the least squares method is significantly higher than that of the standard b_{20} equation, both at different frequencies and tilt angle ranges (Table 5). Therefore, using least squares method fitted equations in the fisheries acoustic assessment of *L. crocea* is clearly superior to the standard b_{20} equation. The b_{20} value in the resulting standard b_{20} expression is smaller than the default b_{20} value of -68 dB for swim bladder fish. Therefore, the use of the standard b_{20} equation in previous surveys would have overestimated the actual TS, which led to a certain underestimation of the resource amount of *L. crocea*. We also suggest using the least square method equation to fit the formula, in order to improve the accuracy of *L. crocea* resource assessment. To obtain the continuous variation pattern of TS with frequency in *L. crocea*, we plotted the spectral curve using Sample 6. The mean TS exhibited a decreasing trend in the frequency range of 18 kHz to 30 kHz, but remained relatively stable between 30 kHz and 300 kHz (Figure 8). A comparison of the spectral curves of the mean TS revealed similar trends in the two tilt angle ranges ($-5°$, $15°$) and ($0°$, $10°$). Gauthier and Horne's [55] study of Pacific herring (*Clupea pallasii*) and Walleye pollock

(*Theragra chalcogramma*) reported similar variations, with TS decreasing at low frequencies and stabilizing at high frequencies. Based on these observations, it is recommended to use frequencies of 70 kHz and 120 kHz in future surveys. In addition, understanding the spectral characteristics of *L. crocea* aids in species identification during mixed-species marine fishery surveys.

Table 5. The least squares method equation and the standard b_{20} equation are established for the two tilt angle ranges under the three acoustic detection frequencies of 70 kHz, 120 kHz, and 200 kHz respectively.

Tilt Angle Distribution	Frequency (kHz)	Least Squares Method Equation	Standard b_{20} Equation
(−5°, 15°)	70	$TS = 32.99 \cdot \log_{10}L - 87.36$ ($R^2 = 0.8686$)	$TS = 20 \cdot \log_{10}L - 70.16$ ($R^2 = 0.6113$)
	120	$TS = 33.26 \cdot \log_{10}L - 87.77$ ($R^2 = 0.9531$)	$TS = 20 \cdot \log_{10}L - 71.23$ ($R^2 = 0.8010$)
	200	$TS = 39.46 \cdot \log_{10}L - 95.51$ ($R^2 = 0.9382$)	$TS = 20 \cdot \log_{10}L - 71.24$ ($R^2 = 0.7096$)
(0°, 10°)	70	$TS = 28.69 \cdot \log_{10}L - 81.71$ ($R^2 = 0.7119$)	$TS = 20 \cdot \log_{10}L - 70.88$ ($R^2 = 0.6453$)
	120	$TS = 32.30 \cdot \log_{10}L - 86.44$ ($R^2 = 0.8460$)	$TS = 20 \cdot \log_{10}L - 71.10$ ($R^2 = 0.7228$)
	200	$TS = 45.87 \cdot \log_{10}L - 103.73$ ($R^2 = 0.9559$)	$TS = 20 \cdot \log_{10}L - 71.48$ ($R^2 = 0.6512$)

Author Contributions: Conceptualization, J.M., S.L. and P.S.; Methodology, J.M., Y.T. and S.L.; Software, J.M. and S.L.; Validation, J.M., X.M. and S.L.; Formal analysis, J.M. and S.L.; Investigation, J.M., L.S., X.M. and S.L.; Resources, L.L.; Data curation, J.M. and L.S.; Writing—original draft, J.M. and S.L.; Writing—review and editing, L.L. and P.S.; Visualization, Y.T.; Supervision, Y.L., X.M. and S.L.; Project administration, L.L.; Funding acquisition, S.L. and P.S. All authors have read and agreed to the published version of the manuscript.

Funding: This study was supported by the Guiding Projects of Fujian Provincial Science and Technology Program of China (2022N0023) and the Scientific Research Foundation of Third Institute of Oceanography, MNR (2019018; 2023019).

Institutional Review Board Statement: The animal study of our manuscript was reviewed and approved by the Ethics Committee of the Third Marine Institute, Ministry of Natural Resources. (Protocol number: TIO-IACUC-06-2024-03-08, Approval date: 8 March 2024).

Data Availability Statement: The datasets generated during and analyzed during the current study are available from the corresponding author upon reasonable request.

Conflicts of Interest: All authors acknowledge that there are no conflicts of interest with their involvement and publication of this work.

References

1. Zhang, H.; Wang, J.; Jing, Y. *Larimichthys crocea* (large yellow croaker): A bibliometric study. *Heliyon* **2024**, *10*, e37393. [CrossRef] [PubMed]
2. Park, D.Y.; Kim, D.G.; Moon, S.Y.; Park, J.H.; Kim, H.Y.; Baeck, G.W. Feeding habits of the large yellow croaker. *Larimichthys crocea in the coastal waters of Jeju Island, Korea. J. Korean Ichthyol.* **2024**, *36*, 182–187.
3. Froese, R.; Pauly, D. *FishBase*; World Wide Web Electronic Publication: Beijing, China, 2014.
4. Wang, L.; Shi, X.; Su, Y.; Meng, Z.; Lin, H. Loss of Genetic Diversity in the Cultured Stocks of the Large Yellow Croaker, Larimichthys crocea, Revealed by Microsatellites. *Int. J. Mol. Sci.* **2012**, *13*, 5584–5597. [CrossRef] [PubMed]
5. Yang, Q.H.; Zhong, C.M.; Zhou, C.; Zhang, Y.; Liao, B.X. Countermeasures and reflections on promoting the quality and safety management of the large yellow croaker (*Larimichthys crocea*) products in Fujian Province. *Fish. Res.* **2020**, *42*, 519–526.
6. Tang, Y. Research advance in fisheries resources assessment by using the acoustic technology in China: A review. *J. Dalian Ocean Univ.* **2023**, *38*, 185–195.
7. Andersen, L.N.; Chu, D.; Heimvoll, H.; Korneliussen, R.; Macaulay, G.J.; Ona, E. Quantitative processing of broadband data as implemented in a scientific split-beam echosounder. *Methods Ecol. Evol.* **2024**, *15*, 317–328. [CrossRef]

8. Kim, H.; Kang, D.; Cho, S.; Kim, M.; Park, J.; Kim, K. Acoustic Target Strength Measurements for Biomass Estimation of Aquaculture Fish, Redlip Mullet (*Chelon haematocheilus*). *Appl. Sci.* **2018**, *8*, 1536. [CrossRef]
9. Zang, J.; Qiu, Y.; Chen, Z.; Zhang, P.; Zhang, K.; Fan, J.; Chen, G.; Cai, Y.; Sun, M. Advances in pelagic fishery resources survey and assessment in open South China Sea. *South China Fish. Sci.* **2018**, *14*, 118–127.
10. Zhang, H.; Yang, D.; Wei, Q.; Du, H.; Zhang, H.; Chen, X. Hydro-Acoustic surver on fishes in the reach from Gezhouba dam to Gulaobei of the Yangtze river. *Resour. Environ. Yangtze Basin* **2007**, *16*, 86–91.
11. Zhang, H.; Wang, C.; Yang, D.; Du, H.; Wei, Q.; Kang, M. Spatial distribution and habitat choice of adult Chinese sturgeon (*Acipenser sinensis Gray*, 1835) downstream of Gezhouba Dam, Yangtze River, China. *J. Appl. Ichthyol.* **2014**, *30*, 1483–1491. [CrossRef]
12. Xie, X.; Zhang, H.; Wang, C.; Wu, J.; Ye, H. Are river protected areas sufficient for fish conservation? Implications from large-scale hydroacoustic surveys in the middle reach of the Yangtze River. *BMC Ecol.* **2019**, *19*, 42–55. [CrossRef] [PubMed]
13. Rudstam, L.G.; Parker-Stetter, S.L.; Sullivan, P.J.; Warner, D.M. Towards a standard operating procedure for fishery acoustic surveys in the Laurentian Great Lakes, North America. *ICES J. Mar. Sci.* **2009**, *66*, 1391–1397. [CrossRef]
14. Wu, Z.; Li, J.; Zhu, S.; Li, X. Seasonal variation of fish density and behavior in Shijiao Reservoir, Beijiang River by using hydroacoustic methods. *J. Fish. Sci. China* **2018**, *25*, 674–681. [CrossRef]
15. Laouar, H.; Djemali, I. Seasonal inter-calibration between acoustic and multi-mesh gillnets sampling for fish biomass assessment in reservoirs. *J. Appl. Ichthyol.* **2018**, *34*, 850–860. [CrossRef]
16. Lyu, S.; Qiu, C.; Xue, M.; Zhu, Z.; Qiu, Y.; Tong, J. Unveiling the Acoustic Signature of *Collichthys lucidus*: Insights from X-ray Morphometry-Informed Acoustic Modeling and Empirical Analysis. *Fishes* **2024**, *9*, 304. [CrossRef]
17. Foote, K.G.; Aglen, A.; Nakken, O. Measurement of fish target strength with a split-beam echo sounder. *J. Acoust. Soc. Am.* **1986**, *80*, 612–621. [CrossRef]
18. Foote, K.G. Importance of the swimbladder in acoustic scattering by fish: A comparison of gadoid and mackerel target strengths. *J. Acoust. Soc. Am.* **1980**, *67*, 2084–2089. [CrossRef]
19. Hazen, E.L.; Horne, J.K. A method for evaluating the effects of biological factors on fish target strength. *ICES J. Mar. Sci.* **2003**, *60*, 555–562. [CrossRef]
20. Jech, J.M.; Horne, J.K.; Chu, D.; Demer, D.A.; Francis, D.T.I.; Gorska, N.; Jones, B.; Lavery, A.C.; Stanton, T.K.; Macaulay, G.J. Comparisons among ten models of acoustic backscattering used in aquatic ecosystem research. *J. Acoust. Soc. Am.* **2015**, *138*, 3742–3764. [CrossRef]
21. Bernasconi, M.; Patel, R.; Nottestad, L.; Pedersen, G.; Brierley, A.S. The effect of depth on the target strength of a humpback whale (*Megaptera novaeangliae*). *J. Acoust. Soc. Am.* **2013**, *134*, 4316–4322. [CrossRef]
22. Khanzhyn, D.; Heuven, W.J.B.V.; Rataj, K. The impact of spatial and verbal working memory load on semantic relatedness judgements. *Bull. Psychon. Soc.* **2024**, *31*, 781–789. [CrossRef] [PubMed]
23. Sobradillo, B.; Boyra, G.; Martinez, U.; Carrera, P.; Pea, M.; Irigoien, X. Target Strength and Swimbladder Morphology of Mueller's Pearlside (*Maurolicus muelleri*). *Sci. Rep.* **2019**, *9*, 17311. [CrossRef] [PubMed]
24. Hasegawa, K.; Yan, N.; Mukai, T. In Situ Broadband Acoustic Measurements of Age-0 Walleye Pollock and Pointhead Flounder in Funka Bay, Hokkaido, Japan. *J. Mar. Sci. Technol.* **2021**, *29*, 135–145. [CrossRef]
25. Henderson, M.J.; Horne, J.K. Comparison of in situ, ex situ, and backscatter model estimates of Pacific hake (*Merluccius productus*) target strength. *Can. J. Fish. Aquat. Sci.* **2007**, *64*, 1781–1794. [CrossRef]
26. Madirolas, A.; Membiela, F.A.; Gonzalez, J.D.; Cabreira, A.G.; Dell'Erba, M.; Prario, I.S.; Blanc, S. Acoustic target strength (TS) of argentine anchovy (*Engraulis anchoita*): The nighttime scattering layer. *ICES J. Mar. Sci.* **2017**, *74*, 1408–1420. [CrossRef]
27. Sawada, K.; Uchikawa, K.; Matsuura, T.; Sugisaki, H.; Amakasu, K.; Abe, K. In Situ and Ex Situ Target Strength Measurement of Mesopelagic Lanternfish, *Diaphus theta* (Family Myctophidae). *J. Mar. Sci. Technol.* **2011**, *19*, 302–311. [CrossRef]
28. Chu, D.; Foote, K.G.; Stanton, T.K. Further analysis of target strength measurements of Antarctic krill at 38 and 120 kHz: Comparison with deformed cylinder model and inference of orientation distribution. *J. Acoust. Soc. Am.* **1993**, *93*, 2985–2988. [CrossRef]
29. Clay, C.S. Composite ray-mode approximations for backscattered sound from gas-filled cylinders and swimbladders. *J. Acoust. Soc. Am.* **1992**, *92*, 2173–2180. [CrossRef]
30. Smith, J.N.; Ressler, P.H.; Warren, J.D. A distorted wave Born approximation target strength model for Bering Sea euphausiids. *ICES J. Mar. Sci.* **2013**, *70*, 204–214. [CrossRef]
31. Lu, D. Researches on Acoustic Scattering of Elastic Target on Finite Element Methods. Master's Thesis, Harbin Engineering University, Harbin, China, 2014.
32. Zhu, Z.; Tong, J.; Xue, M.; Qiu, C.; Lyu, S.; Liu, B. Investigations on Target Strength Estimation Methods: A Case Study of Chub Mackerel (*Scomber japonicus*) in the Northwest Pacific Ocean. *Fishes* **2024**, *9*, 307. [CrossRef]
33. Kang, H.Y.; Lee, D.J. Fish length dependance of acoustic target strength for large yellow croaker. *Korean Sci.* **2003**, *39*, 239–248.
34. Tian, S.; Xue, M.; Tong, J.; Wang, X. A review of the acoustic target strength of fishery resources in the North Pacific Ocean. *J. Fish. Sci. China* **2021**, *28*, 371–379.
35. Foote, K.G. Rather-high-frequency sound scattering by swimbladdered fish. *J. Acoust. Soc. Am.* **1985**, *78*, 688–700. [CrossRef]
36. Yoon, E.; Oh, W.S.; Lee, H.; Hwang, K.; Kim, D.N.; Lee, K. Comparison of Target Strength of Pacific Herring (*Clupea pallasii Valenciennes*, 1847) from Ex-Situ Measurements and a Theoretical Model. *Water* **2021**, *13*, 3009. [CrossRef]

37. Li, C.; Chu, D.; Horne, J.; Li, H. Comparison of coherent to incoherent kirchhoff-ray-mode (KRM) models in predicting backscatter by swim-bladder-bearing fish. *J. Mar. Sci. Eng.* **2023**, *11*, 473. [CrossRef]
38. Li, J. Acoustic Echo Strength of Three Tropical Tuna Species. Master's Thesis, Shanghai Ocean University, Shanghai, China, 2022.
39. Kang, D.; Sadayasu, K.; Mukai, S.K.D.K.K. Target strength estimation of black porgy *Acanthopagrus schlegeli* using acoustic measurements and a scattering model. *Fish. Sci.* **2004**, *70*, 819–828. [CrossRef]
40. Xue, M. Scattering Characteristics of Chub Mackerel (*Scomber japonicus*) in the Northwest Pacific Ocean and the Use in Stock Assessment. Master's Thesis, Shanghai Ocean University, Shanghai, China, 2022.
41. Wu, H.; Hu, Z.; Ge, H.; Gao, Y.; Pan, J.; Liu, Q. Target strength of *Hypophthalmichthys molitrix* and *Aristichthys nobilisbased* on Kirchoff-ray approximation model. *J. Fish. China* **2022**, 1–14. Available online: http://kns.cnki.net/kcms/detail/31.1283.S.20220721.1344.002.html (accessed on 18 October 2024).
42. Luan, Y.; Guan, C.; Shi, X.; Shen, Z. The relationship between acoustic backscattering property and weight and body length of big yellow croaker and black rockfish. *Prog. Fish. Sci.* **2011**, *32*, 47–52.
43. Shang, X. The Research on Acoustical Target Strength of Yellow Croaker, Pseudosciaena Polyactis and Pomfret. Master's Thesis, Zhejiang Ocean University, Zhoushan, China, 2015.
44. Clay, C.S.; Clarence, S. Acoustic models of fish: The Atlantic cod (*Gadus morhua*). *J. Acoust. Soc. Am.* **1994**, *96*, 1661–1668. [CrossRef]
45. Love, R.H. Dorsal-Aspect Target Strength of an Individual Fish. *J. Acoust. Soc. Am.* **1971**, *49*, 816–823. [CrossRef]
46. Rose, G.A.; Porter, D.R. Target-strength studies on Atlantic cod (*Gadus morhua*) in Newfoundland waters. *ICES J. Mar. Sci.* **1996**, *53*, 259–265. [CrossRef]
47. Sun, Y.; Tang, Y.; Xing, B.; Li, H.; Bi, F.; Ma, Z. Target strength of northern whitefish Coregonus peled by using Kirchhoff ray mode methods. *J. Dalian Ocean Univ.* **2021**, *36*, 310–316.
48. Furusawa, M. Prolate spheroidal models for predicting general trends of fish target strength. *J. Acoust. Soc. Jpn.* **1988**, *9*, 13–24. [CrossRef]
49. Lee, S.J.; Lee, Y.W.; Kim, J.I.; Oh, T.Y.; Lee, K.H. Target strength estimation of dominant species in marine ranching ground of Jeju coastal water by KRM model. *J. Korean Soc. Fish. Ocean Technol.* **2010**, *46*, 157–163. [CrossRef]
50. Palermino, A.; Felice, A.D.; Canduci, G.; Biagiotti, I.; Costantini, I.; Centurelli, M.; Leonori, I. Application of an analytical approach to characterize the target strength of ancillary pelagic fish species. *Sci. Rep.* **2023**, *13*, 15182. [CrossRef] [PubMed]
51. Manik, H.M. Quantifying Fish Backscattering using SONAR Instrument and Kirchhoff Ray Mode (KRM) Model. *J. Phys. Conf. Ser.* **2016**, *739*, 012055. [CrossRef]
52. Simón, E.A.; Cellio, J.; Raveau, M.P.; Feuillade, C. Low frequency scattering from dynamic fish schools based on collective animal behavior modeling. *J. Acoust. Soc. Am.* **2015**, *137*, 2361.
53. Ona, E. Physiological factors causing natural variations in acoustic target strength of fish. *J. Mar. Biol. Assoc. UK* **1990**, *70*, 107–127. [CrossRef]
54. Fässler, S.M.M.; Gorska, N.; Ona, E.; Fernandes, P.G. Differences in swimbladder volume between Baltic and Norwegian spring-spawning herring: Consequences for mean target strength. *Fish. Res.* **2007**, *92*, 314–321. [CrossRef]
55. Gauthier, S.; Horne, J.K.; Sciences, A. Acoustic characteristics of forage fish species in the Gulf of Alaska and Bering Sea based on Kirchhoff-approximation models. *Can. J. Fish. Aquat. Sci.* **2004**, *61*, 1839–1850. [CrossRef]

fishes

MDPI

Article

Temporal and Spatial Distribution Characteristics of Fish Resources in a Typical River–Lake Confluence Ecosystem During the Initial Period of Fishing Ban

Huifeng Li [1,2,3], Xujun Yu [4], Bingbing Wu [1,2,3], Lixiong Yu [5], Dengqiang Wang [5], Ke Wang [5], Sheng Wang [6], Daqing Chen [5], Yuefei Li [1,2,3], Xinbin Duan [5] and Jie Li [1,2,3,*]

1 Pearl River Fisheries Research Institute, Chinese Academy of Fishery Sciences, Guangzhou 510380, China; zjlihuifeng@prfri.ac.cn (H.L.)
2 Key Laboratory of Aquatic Animal Immune Technology of Guangdong Province, Guangzhou 510380, China
3 Guangzhou Scientific Observing and Experimental Station of National Fisheries Resources and Environment, Guangzhou 510380, China
4 Beijing Central Ring Geyi Technology Consultation Co., Ltd., Beijing 100041, China
5 Fishery Resources and Environmental Science Experimental Station of the Upper-Middle Reaches of Yangtze River, Ministry of Agriculture and Rural Affairs, Yangtze River Fisheries Research Institute, Chinese Academy of Fishery Science, Wuhan 430223, China; wangdq@yfi.ac.cn (D.W.)
6 Department of agriculture and Rural Affairs of Jiangxi Province, Nanchang 330000, China; wangsheng_86@163.com
* Correspondence: lijie1561@163.com; Tel.: +86-020-8151-6099

Abstract: This study provides a comprehensive analysis of the spatio-temporal dynamics of fish resources in the confluence waters of Poyang Lake and the Yangtze River, focusing on the initial phase of a 10-year fishing ban implemented in January 2020. Through hydroacoustic surveys conducted during both high-water (September 2020) and low-water (January 2021) periods, we identified significant variations in fish density and individual size across different sections. During the high water level period, fish concentrations were primarily observed in the confluence area between the Yangtze River and Poyang Lake, exhibiting higher densities compared to other regions. Conversely, fish congregated in the deep-water zones of the main river during the low water level period. The fish population was dominated by small to medium-sized individuals, with mean body lengths of 12.47 cm and 12.62 cm during the high and low water level periods, respectively. Notably, 42 and 33 fish species were recorded during the high-water and low-water surveys, respectively, emphasizing the region's rich biodiversity. Importantly, the study demonstrates that the fishing ban has resulted in substantial increases in both fish density and mean body length, underscoring its effectiveness in fostering fish population recovery. These findings provide critical baseline data to inform scientific conservation and management strategies in this ecologically sensitive river–lake ecosystem.

Keywords: hydroacoustics; fish distribution; spatial and temporal variation; fishing ban; biodiversity conservation

Key Contribution: This study provides novel insights into the temporal and spatial distribution of fish resources in a river–lake confluence ecosystem during the initial phase of a fishing ban. The findings reveal significant increases in fish density and average body length, demonstrating the ban's effectiveness in fostering fish population recovery and underscoring the importance of conservation efforts in such ecologically sensitive areas.

1. Introduction

The confluence of Poyang Lake and the Yangtze River serves as a transition zone between its midstream and downstream [1,2]. This extensive river segment, spanning approximately 151.9 km in length, exemplifies an abundance of water resources and a diverse array of habitats, thereby creating highly conducive ecological conditions for the proliferation and development of aquatic organisms [3]. Naturally connected to Poyang Lake, the confluence area facilitates the completion of fish life cycles by offering various habitats. The main channel serves as a migration route for essential economic fish species such as the four major Chinese carps, and it is also a feeding ground for most settled cyprinid fish species [4,5], and seasonal flooding results in distinct fish distribution patterns between the main channel and the Yangtze River main stream [6]. However, recent years have witnessed increasing human water-related activities, leading to a deteriorating situation of fish resources in the confluence area, which connects the main channel of Poyang Lake and the Yangtze River. These issues include a decline in fish species diversity, a shift towards smaller and younger individuals, and homogenization of the fish community structure [7,8].

In January 2020, the Ministry of Agriculture and Rural Affairs announced the implementation of a "10-year fishing ban" in key water areas of the Yangtze River basin. Rapidly assessing the distribution of fishery resources under this comprehensive fishing ban has become a current research focus. Currently, there is a dearth of research utilizing acoustic techniques to evaluate fish resources in the confluence of Poyang Lake and the Yangtze River. There are only a few studies evaluating the fish resources in a single habitat of Poyang Lake or the Yangtze River, such as the significant recovery of *Coilia nasus* populations in the Poyang Lake after the fishing ban, and the improvement of miniaturization trends [9,10]. With traditional survey methods facing limitations, acoustic assessment techniques have gradually been adopted in the main stream of the Yangtze River and lakes [11,12]. These methods enable rapid and accurate monitoring of fishery resources without damaging the fish population. Against this backdrop, this study conducted acoustic surveys in the main stream of the Yangtze River in the Jiujiang section and the connecting channel in the Poyang Lake during the initial stages of the fishing ban in September 2020 and January 2021. These surveys analyzed the distribution characteristics of fish resources during both the high and low water level period, aiming to provide baseline data for the implementation of the fishing ban policy and inform scientific conservation and management efforts for fish resources in the confluence of the Poyang Lake and the Yangtze River.

2. Materials and Methods

2.1. Study Area and Procedures

The Jiujiang section of the Yangtze River spans from Fuxingzhou sandbar to Mianchuanzhou sandbar, with a total length of approximately 151.9 km. This section serves as a transition zone between the middle and lower reaches of the Yangtze River. The confluence is naturally connected to Poyang Lake and the Yangtze River, forming a complex ecosystem that integrates rivers and lakes [13]. For the purpose of comprehensively analyzing the temporal and spatial distribution characteristics of fish resources in the confluence of Poyang Lake and the Yangtze River, we divided the hydroacoustic detection area into three distinct regions based on the characteristics of the river sections. Section A covers the stretch from Fuxingzhou sandbar to the entrance of Poyang Lake in the main stream of the Yangtze River, spanning approximately 50 km. Section B extends from the Zhangjiazhou sandbar to the Mianchuan sandbar, with a length of roughly 56 km. Lastly, Section C encompasses the area from the entrance of Poyang Lake in Hukou County to Pingfeng waters, covering approximately 30 km (Figure 1).

In each sampling section, catches were obtained by fishing with various nets. Each evening (18:00–06:00), fixed gillnets that were 20 mm in size were deployed between 15 and 30 m offshore. During the daytime, from 6:00 to 18:00, drift nets that were 50 and 70 mm in size were employed in the deep-water area. Hydroacoustic data were collected, combined with the catch surveying in various sections with different niches (such as backwaters, pools, etc.) using different types of nets. The phylogenetic categorization of fish was completed as Chen [14]. Anesthesia was administered to the fish using MS-222 (Sigma Aldrich Chemical Co., St. Louis, USA), and each species' body length was measured in millimeters. Fish species safeguarded by the Chinese government and those threatened species on the IUCN (International Union for the Conservation of Nature and Natural Resources) and CITES (Convention on International Trade in Endangered Species of Wild Fauna and Flora) lists were reintroduced back into the water.

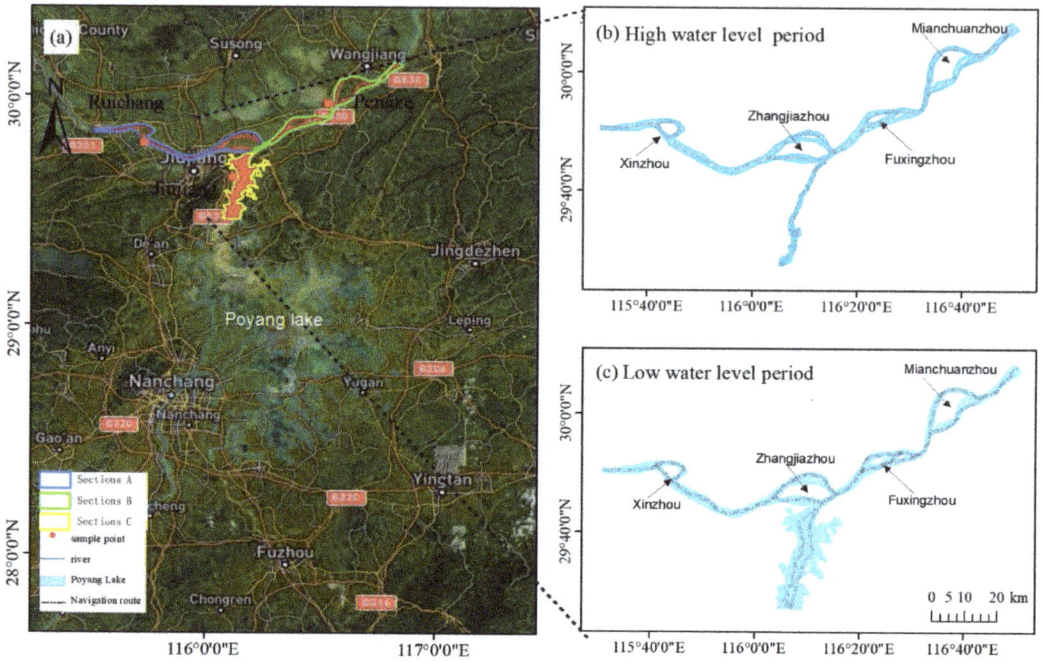

Figure 1. (**a**) Partition and sampling point distribution. (**b**) Schematic diagram of hydroacoustic detection area and route in high water level period. (**c**) Schematic diagram of hydroacoustic detection area and route in low water level period.

In this study, we conducted hydroacoustic surveys during September 2020 and January 2021, along with two different water periods (Figure 1) in the Yangtze River basin using a calibrated SIMRAD EY60 system. The system's working power was 300 W, its pulse width was 64 μs, and its transducer frequency was 200 kHz with an angle of 7° at −3 dB. The transducer was installed on the front of the boat at a depth of 0.5 m under the water surface, with a navigation speed of 8–10 km/h. The GPS data were gathered using a GPS receiver (Garmin, Taiwan, China), and the instrument was calibrated using a tungsten–copper metal ball with a diameter of 13.7 mm before starting the investigation [15]. Throughout the detection process, the TS threshold was adjusted to −70 dB, representing the minimum setting that effectively minimized noise interference in the echogram. The pulse interval was 5 ping/s. Through observation of the echogram, it was found that the fish density in the survey area was low and most individual echoes did not overlap, thus the echo counting method was adopted. The other parameters are referenced in Supplementary Table S1. All detection activities were conducted during daylight hours, specifically, from 8:30 to 16:30. The degree of coverage values for each study was determined using Formula (1) [16], in which Λ is the coverage value, L represents the total length of acoustic transects, and A is the studied area. The results varied from 12 to 15. These coverage values were above the minimum recommendations noted in the literature [17,18].

$$\Lambda = D/\sqrt{A} \quad (1)$$

2.2. Data Processing and Transformation

The dataset incorporated for analysis encompassed the range from 1 m beneath the water surface to 0.5 m above the riverbed. Before processing the data, we first made a judgment based on the overall acoustic characteristics of the fish. We observed that the fish density in the survey area was low, and most individual echoes did not overlap. Therefore, we decided to use the echo counting method in our study. A comprehensive four-step procedure was implemented to identify targets [17]. Initially, file transformation was executed, where raw data files (.raw) were converted into the .uuu format using the converter tool within Sonar5-Pro software (Kongsberg Maritime Inc., Horten, Norway) [18]. Subsequently, bottom detection was carried out using an image analysis detector to delineate the riverbed in each file. Manual adjustments were then made to refine the detected riverbed line. In the third stage, target tracking was performed to detect individual targets, with optimal parameter configurations applied. Lastly, track filtering was applied to the acquired fish-basket dataset to sieve out targets based on specific criteria, and the targets with No. Echoes greater than 4, Max Ping Gap equals 2 pings, and Gating Range equals 0.3 m. The population density of mature individuals was subsequently estimated using the following formula:

$$\text{Density} = \text{target number}/\text{survey volume}. \tag{2}$$

$$\text{Survey volume} = 0.5 \times (2 \times H \times \tan 3.5^{\circ} \times H) \times \text{sampling pings} \times 0.42, \tag{3}$$

$$TS = 25.76 \text{ Log } TL - 105.32, \tag{4}$$

According to the calculation empirical Formula (2), the number of the target was acquired through the aforementioned procedure. The quantity of water in the investigation was determined by modeling it as a triangular prism, employing the subsequent mathematical expression. According to the calculation empirical Formula (3), where H denotes the mean water depth throughout the sampling duration, the constant value of 0.42 has been derived from the mean vessel velocity of 9 km per hour, signifying the average distance traversed by the vessel in response to a single ping. Since the fish population in Poyang Lake is predominantly composed of carps, we employ the carp-specific regression formula developed by Frouzova et al. as Formula (4) [19], where TL denotes the total length in millimeters.

In this study, the survey routes during both the high and low water level periods were largely consistent, with each being divided into 1 km segments. The fish density in each segment was calculated, and these density values along with their respective coordinate data were imported into ArcGIS. The Inverse Distance Weighting (IDW) method was then used for interpolation to generate a horizontal distribution map of fish density in this river section [20]. When analyzing the spatial distribution characteristics of fish, non-parametric tests were conducted using SPSS to analyze the target strength and density of fish in different river sections and periods.

3. Results

3.1. Fish Target Strength and Spatial Distribution Characteristics

The average target strength (TS) of fish during the high water level stage is −51.32 dB, corresponding to an average total length of 12.47 cm. There are significant differences in TS values between different sections ($p < 0.05$), with a decreasing trend from section C to section A. In contrast, during low water levels, the average TS is slightly different, at −51.18 dB, corresponding to an average total length of 12.62 cm. The TS value of section A is significantly lower than that of sections B and C ($p < 0.05$) (Figure 2). During these two periods, the TS values were concentrated between −70 and −43 dB, with a proportion greater than 90%, indicating that small and medium-sized fish were dominant. The proportions in the five intervals of −37 to −34, −40 to −37, and −43 to −40 are all less than 5% (Figure 3). These findings indicate that the overall size distribution of fish remains similar under different water level conditions; however, the distribution proportion of fish varies within a specific target strength range. Especially, the proportion of the maximum fish population recorded during the two water levels varies within different TS ranges, with the maximum proportion occurring between −67 and −64 dB during low water levels and −52 to −49 dB during high water levels.

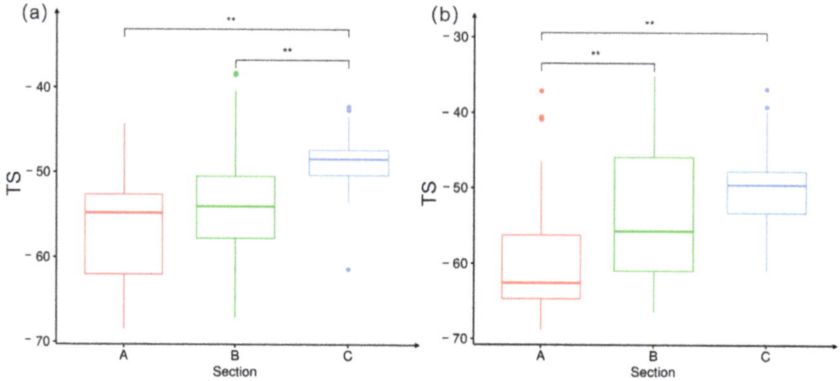

Figure 2. Distribution of the target strengths between different sections during the high (**a**) and low (**b**) water level periods. (The box plot illustrates the following statistical measures: the minimum value, the first quartile–25th percentile, the median, the third quartile–75th percentile, and the maximum value. Outliers are marked with points. The horizontal lines connecting the boxes indicate a comparison between the two river sections during these periods. The asterisks above these lines signify the level of significance in the differences observed; ** denotes a highly significant difference with $p < 0.01$.)

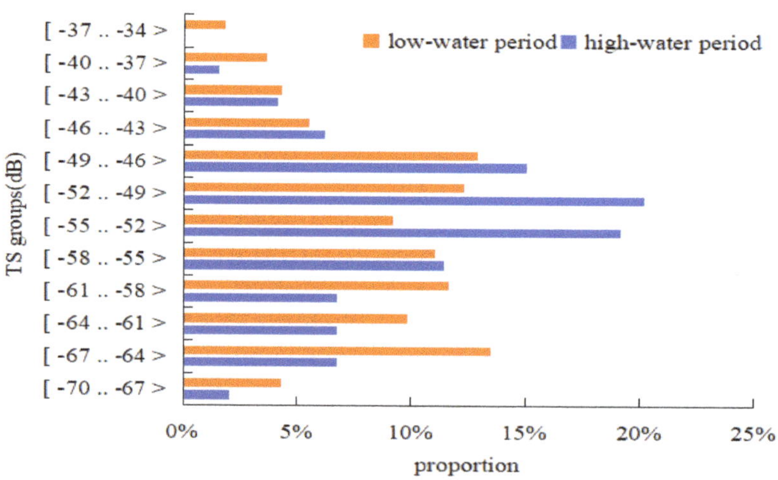

Figure 3. Distribution of different groups of target strength during the high and low water level periods.

Based on the total length of fish, they were categorized into four groups: 0–10 cm, 10–20 cm, 20–30 cm, and >30 cm (Figure 4). The analysis revealed that fish within the 0–10 cm length group dominated the study area. During high water levels, the proportions of 0–10 cm fish in sections A, B, and C were 58.06%, 51.32%, and 4.00%, respectively. Notably, section C exhibited higher proportions of fish in the 10–20 cm and 20–30 cm categories compared to sections A and B. Among fish exceeding 30 cm in length, section B had the highest proportion. During low water levels, 0–10 cm fish were primarily distributed in sections A and B, while 10–20 cm fish were concentrated in section C. Fish in the 20–30 cm and >30 cm categories were mainly found in sections B and C. As illustrated in Figure 5, during high water levels, larger fish were primarily concentrated in the upstream area approximately 15 km from Zhangjiazhou to Fuxingzhou in sections C and B. In contrast, during low water levels, they were mainly distributed near the confluence of sections C and B, specifically, at the tail of the sandbar in Zhangjiazhou and the head of the sandbar in Mianchuanzhou.

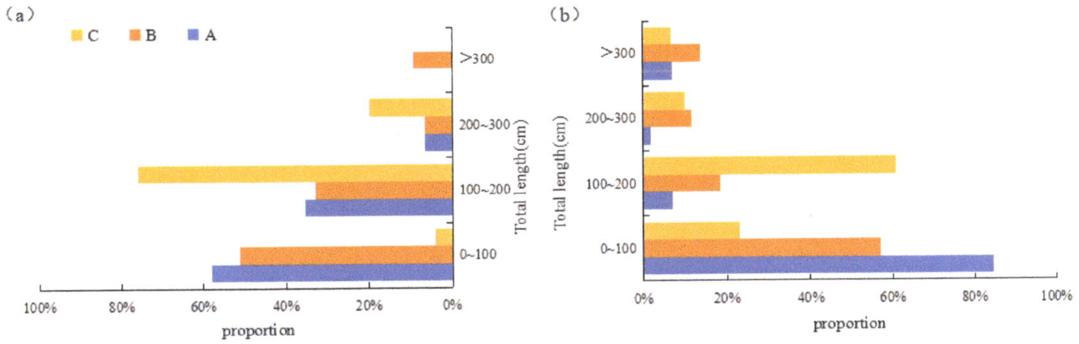

Figure 4. Distribution of total length during the high (**a**) and low (**b**) water level periods (A, B, and C are the codes for different river sections, as shown in Figure 1).

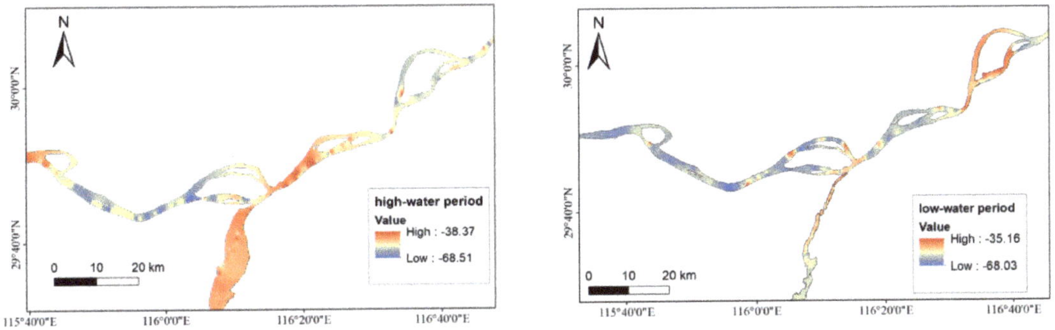

Figure 5. Distribution map of the fish target strength during the high and low water level periods.

3.2. Spatial Distribution Characteristics of Fish Density

The depth and fish density in the study area varied spatially and temporally (Figure 6). As shown in Figure 6a, the average water depth in the autumn study area is 16.72 ± 4.37 m, with a maximum water depth of 39.12 m and a minimum water depth of 4.3 m. The average water depths of river sections A, B, and C are 16.72 ± 4.37 m, 22.40 ± 7.38 m, and 10.20 ± 2.72 m, respectively. The average water depth in the winter research area is 10.71 ± 4.99 m, with a maximum depth of 33.49 m and a minimum depth of 2.65 m. The average water depths of river sections A, B, and C are 9.33 ± 5.56 m, 13.36 ± 5.97 m, and 10.11 ± 2.23 m, respectively. During the high water level period (September 2020), the average fish density was 93.85 ind./1000 m^3, ranging from 0 to 680.61 ind./1000 m^3. The fish density in sections A, B, and C were 78.62 ind./1000 m^3, 96.93 ind./1000 m^3, and 106.78 ind./1000 m^3, respectively (Figure 6b). There were no significant differences in fish density among the three sections ($p > 0.05$). The horizontal distribution map of fish density in Figure 7 showed a patchy aggregation pattern, mainly concentrated from the tail of the sandbar in Zhangjiazhou to the 15 km downstream of the confluence of Poyang Lake with the main stream, with scattered distributions at the head of the sandbar in Mianchuanzhou and the tail of the sandbar in Fuxingzhou.

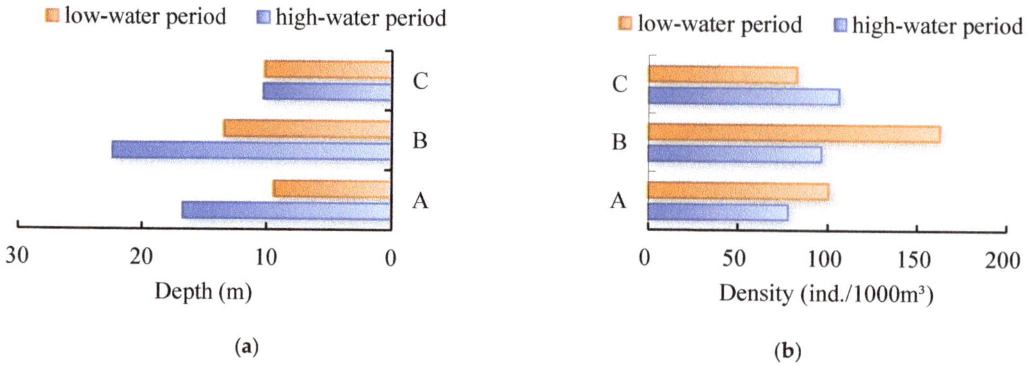

Figure 6. The water depth (**a**) and fish density (**b**) in different sections during the high and low water level period (A, B, and C are the codes for different river sections, as shown in Figure 1).

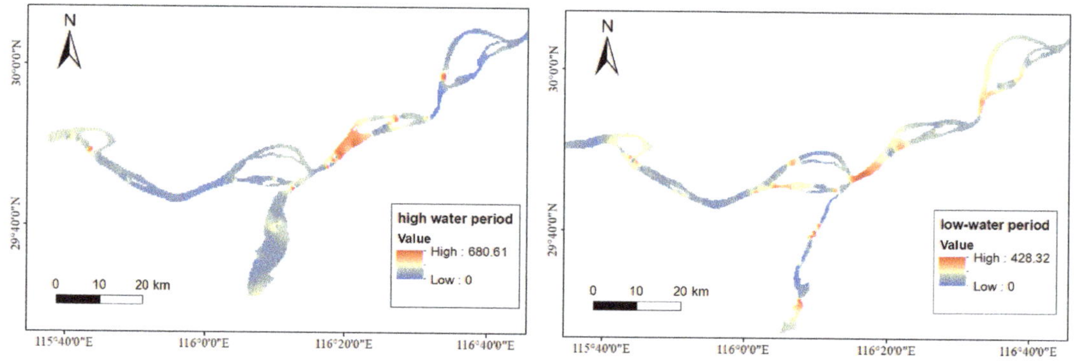

Figure 7. Distribution map of fish density during the high and low water level period.

The average fish density in the study area during the low water level period was 111.17 ind./1000 m^3, ranging from 0 to 428.32 ind./1000 m^3. The fish density in sections A, B, and C were 101.0 ind./1000 m^3, 163.21 ind./1000 m^3, and 83.29 ind./1000 m^3, respectively. The fish density in section B was significantly greater than that in sections A and C ($p < 0.05$). The horizontal distribution map of fish density (Figure 7) showed that during the low water level period, fish were mainly distributed from the front and tail of the sandbar in Zhangjiazhou to the 15 km downstream of the confluence of Poyang Lake with the main stream, with scattered distributions at Pingfeng Mountain in the Poyang Lake, the head of the sandbar in Mianchuanzhou, and the tail of the sandbar in Fuxingzhou.

3.3. Composition of Fish Population

During the high water level period of the study area (September 2020), a total of 42 species were collected, belonging to four orders, seven families, and 26 genera. As shown in Table 1, the dominant species are *Megalobrama terminalis, Coilia brachygnathus, Carassius auratus*, and *Hypophthalmichthys molitrix*. The fish species with a quantity proportion greater than 1% included *Coilia brachygnathus, Megalobrama terminalis, Acheilognathus macropterus, Carassius auratus, Siniperca chuatsi, Hypophthalmichthys molitrix, Saurogobio dabryi, Culter alburnus, Saurogobio dumerili, Culter dabryi, Pelteobaggrus nitidus, Parabramis pekinensis, Megalobrama amblycephala, Aristichthys nobilis*, and *Xenocypris davidi*. The body length of the fish caught in this river section ranged from 3.60 to 83.80 cm, with the arithmetic mean of the body length 20.25 cm.

Table 1. The main characteristics of fish population ratio and body length during the two water periods.

High Water Level Period				Low Water Level Period			
Species	Proportion (%)	Length Range (cm)	Average Length (cm)	Species	Proportion (%)	Length Range (cm)	Average Length (cm)
Coilia brachygnathus	20.04%	11.30~35.40	23.44	*Coilia brachygnathus*	16.06%	9.50~35.00	21.59
Megalobrama terminalis	13.89%	10.70~37.40	24.10	*Culter alburnus*	9.28%	11.00~30.80	21.47
Acheilognathus macropterus	11.61%	3.60~10.00	6.11	*Pseudobrama simoni*	8.72%	9.00~14.00	11.63
Carassius auratus	8.36%	4.50~28.80	19.09	*Parabramis pekinensis*	6.94%	8.30~39.70	20.70
Siniperca chuatsi	4.70%	8.30~48.30	21.78	*Hypophthalmichthys molitrix*	6.78%	13.50~49.20	26.91
Hypophthalmichthys molitrix	4.08%	17.00~83.80	34.60	*Acheilognathus macropterus*	6.54%	2.00~11.50	8.08
Saurogobio dabryi	3.73%	6.40~16.60	10.05	*Siniperca chuatsi*	6.38%	10.50~43.30	21.38
Culter alburnus	3.52%	9.30~47.50	20.33	*Carassius auratus*	6.05%	7.40~28.00	15.95
Saurogobio dumerili	3.32%	5.40~13.50	8.42	*Pelteobagrus fulvidraco*	5.33%	9.00~35.00	13.46
Culter dabryi	3.04%	6.70~35.20	22.75	*Siniperca kneri*	3.87%	14.40~30.00	20.64
Pelteobaggrus nitidus	2.49%	5.80~24.20	13.76	*Pelteobagrus fulvidraco*	3.79%	8.90~22.00	12.29
Parabramis pekinensis	2.49%	6.40~39.20	26.65	*Megalobrama terminalis*	3.71%	10.90~43.00	19.62
Megalobrama amblycephala	2.28%	9.70~34.30	27.17	*Xenocypris argentea*	2.18%	11.50~21.70	15.70
Aristichthys nobilis	2.00%	16.50~61.00	39.62				
Xenocypris davidi	1.80%	7.70~20.80	17.18				

During the low water level period (January 2021), 33 species of fish were caught, belonging to four orders, six families, and 22 genera. As shown in Table 1, he dominant species were *Hypophthalmichthys molitrix, Coilia brachygnathus, Parabramis pekinensis,* and *Siniperca chuatsi*. The fish species with a quantity proportion greater than 1% included *Coilia brachygnathus, Culter alburnus, Pseudobrama simoni, Parabramis pekinensis, Hypophthalmichthys molitrix, Acheilognathus macropterus, Siniperca chuatsi, Carassius auratus, Pelteobagrus fulvidraco, Siniperca kneri, Pelteobagrus fulvidraco, Megalobrama terminalis,* and *Xenocypris argentea*. The body length of the fish caught in this river section during the low water level period ranged from 2.00 to 49.20 cm, with the arithmetic mean of the body length 18.56 cm.

4. Discussion

4.1. Spatial and Temporal Characteristics of Fish Density

During the low water level period, the average density of fish in the C area is significantly lower than that in the main stream areas A and B, with the B area of the main stream having the highest average fish density. During the high water level period, the average density of fish is highest in the C section. These changes may be related to seasonal migratory behaviors of fish such as foraging, fattening, and overwintering. During the high water level period, shallow beaches, sandbars, and vegetation near the shore are submerged in the Poyang Lake, providing complex habitats for fish [21,22]. The river channel connecting the lakes in the C section has also become a feeding and breeding ground for various fish, especially a busy migration channel [23]. When the water level drops, fish gather in the deep waters of the river for overwintering [24]. The main stream has more deep-water areas [25,26], and the average water depth in the B section is greater than that in the A section and C section, as described in Section 3.2; this may be the reason for the highest density in the B River section, resulting in the highest density of the B section of the river during the low water level period.

The density distribution of the two water level periods results indicate that the fish are mainly distributed in the Pingfeng Mountain of section C and 15 km below the confluence of section B, with obvious regional characteristics. At the same time, the distribution of fish also shows a characteristic of being close to sandbars, such as the head and tail of sandbars named Mianchuanzhou, Zhangjiazhou, and Fuxingzhou, which are fish gathering areas (Figure 7). As a typical habitat in rivers, sandbars have formed diverse fish habitats due to their complex hydrological environment [27]. At the same time, seasonal water will submerge sandbars, and the submerged plants in the bands on both sides of the sandbars will become habitats and breeding grounds for fish [28].

The Zhangjiazhou area, where fish are mainly distributed, is the only estuary where Poyang Lake flows into the Yangtze River. The ecological ecotone theory suggests that an important feature of this area is high habitat heterogeneity, with high biodiversity and productivity, and the river confluence is beneficial for increasing fish diversity [29,30]. For the sandbar area, seasonal floods inundate the floodplain, providing abundant habitats for fish to inhabit. Due to the temperature gradient and eddies formed at the confluence, nutrients, woody debris, and organic matter gather there, which is conducive to buoyancy and plant growth, providing a rich source of food for fish [31]. As the only intersection connecting the Yangtze River and Poyang Lake, this area can also be seen as an ecological transition zone [7,32]. The continuity and habitat heterogeneity in the time and space of the rivers and lakes are also important reasons for the high density of fish in this area.

4.2. Fish Target Strength

After comparing the average water depth, average TS, and average fish density of three river sections in different water periods, we found that section B has the deepest water depth, followed by section A, and then section C. The average TS value and average density value of fish during high water level periods are lower than those during low water level periods, and the same applies to sections A and B. However, the average TS value and average density value of section C during the high water level are lower than those during low water levels. Further research has found that the distribution pattern is related to physiological habits such as feeding, wintering, and reproduction in fish [24]. Sections A and B are in the Yangtze River, while section C is contained in Poyang Lake. During the low water level period, the water area of the Poyang Lake shrinks, and the C River section becomes the deepest part of the Poyang Lake, making it more suitable for large fish to overwinter here, thus becoming a wintering ground for the fish in the Poyang Lake [33]. The relevant topographic studies conducted in the Poyang Lake area have shown that the water depth of the north channel area (section C) in the Poyang Lake is greater than that in the main lake area and the southern dish lake area [34]. According to the fish catch survey, the main species of fish caught in the C River section are dominated by large, long-bodied fish such as grass carp, silver carp, catfish [35].

The results in Section 3.1 show that the fish assemblage in the study area is predominantly composed of small fish species, with sizes ranging from 1 to 10 cm, accounting for over 45% across all three river sections. This is in accordance with findings reported in previous studies [36,37]. Upon analyzing additional research in Section 3.3, it is evident that the dominant species in this aquatic environment include *Megalobrama terminalis, Coilia brachygnathus, Carassius auratus, Hypophthalmichthys molitrix, Parabramis pekinensis,* and *Siniperca chuatsi,* all of which are primarily small fish species. This aligns with the situation depicted in the results of the current hydroacoustic survey. Furthermore, scholarly research has indicated that the average total length of fish caught may be overestimated due to the employment of nets with larger mesh sizes in fishing surveys resulting in inadequate capture of small fish, thereby biasing the catch towards larger individuals [38]. Consequently, future research endeavors should involve the selection of more suitable nets, thereby improving the consistency with the results of underwater acoustic investigations.

4.3. Changes in Fish Resources Before and After the Fishing Ban

A comparison with research results from the waterway connecting Poyang Lake to the Yangtze River prior to the fishing ban (in 2014) [39] reveals that both the average fish density and Target Strength (TS) values are higher after the implementation of the fishing ban (in September 2020). Specifically, the mean TS value of fish after the ban (−48.23 dB) is greater than before (−56.4 dB). Additionally, the proportion of individuals with TS values ranging from −70 to −55 dB (42.59%) is lower after the ban than before (54.6%), while the proportion of individuals with TS values greater than −40 dB (1.23%) is higher than before (0.12%). The average fish density was 106.78 ind./1000 m^3, which is higher than the 53.7 ind./1000 m^3 before the fishing ban.

These findings suggest that the fishing ban policy has played a positive role in the restoration of fish resources in the Yangtze River basin, with increases in both fish density and individual size, and a mitigation of the trend towards miniaturization in the fish population structure. Reports from other researchers also found the positive effects of the fishing ban policy. For instance, fish in the different sections (Yichang to Chenglingji section, main stream of the upper Yangtze River, Xiangjiaba Reservoir in the lower reaches of Jinsha River, the Dongting Lake) of the Yangtze River basin [40–43] have shown an increasing trend in size after the ban, and both the number of fish species and diversity indices have increased. These observations further illustrate the beneficial impact of the ten-year fishing ban policy on fish resource conservation.

The implementation of the fishing ban policy, as a pivotal initiative aimed at protecting and restoring aquatic ecosystems, has gradually manifested its profound impact [44]. This policy has not only markedly reduced the fishing intensity targeting adult fish populations, thereby providing invaluable space and time for the natural growth of young fish and subsequently promoting a dual increase in the quantity and size of fish resources, but it has also effectively improved fish habitats through a series of scientific and reasonable environmental protection measures and stringent regulation of human activities. These measures encompass, but are not limited to, water purification, ecological restoration, and the rational planning of fishing activities, all of which collectively act upon the fish ecosystem to significantly enrich the food sources for fish and create favorable conditions for the regeneration and proliferation of fish populations.

However, the restoration of ecosystems is a complex and lengthy process, a fact that is particularly evident in the recovery of fish resources within the two key lakes of the Yangtze River basin: Poyang Lake and Dongting Lake. According to relevant research, achieving a basic balance in fish resources in these two lakes is expected to take a time span of 3 to 5 years [45]. Taking Poyang Lake as an example, despite some progress in fish resource recovery in recent years, the overall situation is still far from full restoration. The current level of recovery, to some extent, is roughly comparable to the state in the 1990s [46], indicating that the road to ecosystem restoration remains long and arduous. More severely, the ecosystem of Poyang Lake remains relatively fragile, with limited resistance to external disturbances.

Therefore, the continuous monitoring and evaluation of fish resource recovery in Poyang Lake are of paramount importance. This necessitates maintaining the continuity and stability of the existing fishing ban policy while flexibly adjusting the specific implementation details of the ban measures based on the actual situation of fish resource recovery and the dynamic characteristics of ecosystem changes. By establishing a comprehensive monitoring system, timely access to firsthand data on fish resource recovery can be obtained, providing robust support for scientific decision making. This, in turn, ensures that the fishing ban policy can more effectively promote the protection and restoration of fish resources, contributing to the ecological security and sustainable development of the Yangtze River basin.

5. Conclusions

We utilized hydroacoustic methodologies to undertake a comprehensive and meticulous examination of the spatial distribution patterns of fish in the confluence zone of the Poyang Lake and the Yangtze River. The primary aim of this investigation was to assess the condition of fish resources in this pivotal aquatic ecosystem during the early stages of the implemented fishing ban. With this survey, we intended to generate robust data that could facilitate subsequent scientific evaluations of the fishing ban's efficacy and inform effective conservation strategies for fish resources.

Our findings reveal that fish resources in our study area are predominantly clustered within the confluence region and adjacent sandbar waters. This observation carries substantial implications for comprehending fish behavioral patterns and habitat preferences within distinct ecological settings. Following the rigorous enforcement of the fishing ban policy, we documented pronounced alterations in the fish resource status. A comparative analysis of post-ban survey data with prior reports disclosed an increase in both the average fish density and the average target strength (TS). This shift indicates that the fishing ban policy has exerted a favorable influence on reversing the trend of fish population downsizing and fostering an elevation in the abundance of larger fish species, thereby substantially contributing to the recovery of fish resources in the typical river–lake confluence area.

Nonetheless, this study constitutes a preliminary foray into understanding the impact of the fishing ban policy on fish resources in Poyang Lake. In the future, it is acknowledged that the estimation of fish total length from TS remains an approximation, as highlighted by the extensive hydroacoustic literature. To enhance the accuracy of this estimation, exploring and incorporating various existing equations that consider not only the diverse shapes of fish and the size of the swim bladder but also the positional relationship between the fish and the transducer will be crucial. This will necessitate a more comprehensive and nuanced approach to data analysis, potentially involving advanced modeling and simulation techniques. By addressing these complexities, future research endeavors aim to refine the understanding and application of hydroacoustic methods in fish resource assessment, ultimately leading to more precise and reliable estimates of fish size and abundance.

Supplementary Materials: The following supporting information can be downloaded at: https://www.mdpi.com/article/10.3390/fishes9120492/s1, Figure S1: Pictures of field investigation. (a) Lake scenery. (b) Main fish species. (c) Fish collection. (d) Hydroacoustic survey. (e) Data processing. Table S1: Main value setting of parameters.

Author Contributions: Conceptualization, J.L.; Data curation, X.Y., K.W., and J.L.; Formal analysis, H.L. and X.Y.; Funding acquisition, X.D.; Investigation, X.Y.; Methodology, B.W.; Project administration, J.L.; Resources, D.W.; Software, H.L., B.W., and Y.L.; Supervision, S.W.; Validation, L.Y.; Visualization, D.C.; Writing—original draft, H.L. and X.Y.; Writing—review and editing, H.L. and B.W. All authors have read and agreed to the published version of the manuscript.

Funding: This work was supported by Project of innovation team of survey and assessment of the Pearl River fishery resources (2023TD10); The National Key R&D Program of China (2018YFD0900801); The Project of Yangtze Fisheries Resources and Environment Investigation from the MARA, China; and Chinese Three Gorges Corporation (No: 202003229).

Institutional Review Board Statement: In our study, ethical approval was not required as our fishing activities were solely for the purpose of classification and immediate release back into the water, without any dissection or harm to the fish.

Informed Consent Statement: Informed consent was obtained from all subjects involved in the study.

Data Availability Statement: Data are contained within the article and supplementary materials.

Conflicts of Interest: Author X.Y. is employed by the company Beijing Central Ring Guoji Technology Consultation Co., Ltd. The remaining authors declare that the research was conducted in the absence of any commercial or financial relationships that could be construed as potential conflicts of interest.

References

1. Dronova, I.; Beissinger, S.R.; Burnham, J.W.; Gong, P. Landscape-Level Associations of Wintering Waterbird Diversity and Abundance from Remotely Sensed Wetland Characteristics of Poyang Lake. *Remote Sens.* **2016**, *8*, 462. [CrossRef]
2. Xu, G.; Qin, Z. Flood Estimation Methods for Poyang Lake Area. *J. Lake Sci.* **1998**, *10*, 31–36. [CrossRef]
3. Feng, L.; Hu, C.; Chen, X.; Li, R.; Tian, L.; Murch, B. MODIS observations of the bottom topography and its inter-annual variability of Poyang Lake. *Remote Sens. Environ.* **2011**, *115*, 2729–2741. [CrossRef]
4. Li, Y.; Qian, F.; Silbernagel, J.; Larson, H. Community structure, abundance variation and population trends of waterbirds in relation to water level fluctuation in Poyang Lake. *J. Great Lakes Res.* **2019**, *45*, 976–985. [CrossRef]
5. Huang, L.; Wu, Z.; Li, J. Fish fauna, biogeography and conservation of freshwater fish in Poyang Lake Basin, China. *Environ. Biol. Fishes* **2013**, *96*, 1229–1243. [CrossRef]

6. Li, Y.; Zhang, Q.; Cai, Y.; Tan, Z.; Wu, H.; Liu, X.; Yao, J. Hydrodynamic investigation of surface hydrological connectivity and its effects on the water quality of seasonal lakes: Insights from a complex floodplain setting (Poyang Lake, China). *Sci. Total Environ.* **2019**, *660*, 245–259. [CrossRef] [PubMed]
7. Hu, M. *Analysis of Water Level and Water Environmental Characteristics of Poyang Lake and Its Impact on Fish Community and Migration*; Nanchang University: Nanchang, China, 2009; pp. 56–89.
8. Wang, S.; Duan, X.; Chen, W. Status and changes of fish resources in the Hukou area of Poyang Lake. *Freshw. Fish.* **2016**, *46*, 50–55.
9. Tao, J.; Yang, J.; Xuan, Z. Preliminary Report on the Effects of Resource Recovery on Anadromous Coilia nasus in Poyang Lake Under the National 10-Year Fishing Ban. *Process. Fish. Sci.* **2022**, *43*, 7. [CrossRef]
10. Dong, F.; Zhang, H.; Wei, Q. Protection and development after the ten-year fishing ban in the Yangtze River. *J. Fish.* **2023**, *47*, 029318. [CrossRef]
11. Lian, Y.; Huan, G.; Godlewska, M.; Li, C.; Zhao, X.; Ye, S.; Li, Z. Hydroacoustic assessment of spatio-temporal distribution and abundance of fish resources in the Xiangxi river. *Acta Hydrobiol. Sin.* **2015**, *39*, 920–929. [CrossRef]
12. Tao, J.; Gong, Y.; Tan, X. Spatiotemporal patterns of the fish assemblages downstream of the Gezhouba Dam on the Yangtze River. *Sci. Sin. Vitae* **2012**, *42*, 677–688. [CrossRef]
13. Yang, G.; Zhang, Q.; Wan, R.; Lai, X.; Jiang, X.; Li, L.; Dai, H.; Lei, G.; Chen, J.; Lu, Y. Lake hydrology, water quality and ecology impacts of altered river–lake interactions: Advances in research on the middle Yangtze River. *Hydrol. Res.* **2016**, *47* (Suppl. 1), 1–7. [CrossRef]
14. Chen, Y. *Fauna Sinica. Osteichthyes, Cypriniformes II*; Science Press: Beijing, China, 1998. (In Chinese)
15. Foote, K.G.; Knudsen, H.P.; Vestnes, G.; MacLennan, E.J.; Simmonds, E.J. Calibration of acoustic instruments for fish density estimation: A practical guide. *ICES Coop. Res. Rep.* **1987**, *144*, 84. [CrossRef]
16. Aglen, A. Random Errors of Acoustic Fish Abundance Estimates in Relation to The Survey Grid Density Applied. *Fao Fish. Report.* **1983**, *300*, 293–298.
17. Zhang, H.; Yang, D.; Wei, Q.; Du, H.; Wang, C. Spatial distribution and spawning stock estimates for adult Chinese sturgeon (*Acipenser sinensis* Gray, 1835) around the only remaining spawning ground during the trial operation of the newly constructed Three Gorges Project in the Yangtze River. *J. Appl. Ichthyol.* **2013**, *29*, 1436–1440. [CrossRef]
18. Balk, H. *Sonar4 and Sonar5-Pro Post Processing Systems: Operator Manual*, version 6.0.4; Lindem Data Acquisition A/S: Oslo, Norway, 2013. Available online: http://folk.uio.no/hbalk/sonar4_5/ (accessed on 6 October 2018).
19. Frouzova, J.; Kubecka, J.; Balk, H.; Frouz, J. Target strength of some European fish species and its dependence on fish body parameters. *Fish Res.* **2005**, *75*, 86–96. [CrossRef]
20. Gerasimov, Y.V.; Pavlov , D.D.; Strelnikova , A.P.; Shlyapkin, I.V.; Borisenko, E.S. Impacts of the High-Pressure Bratsk Hydroelectric Power Station on Fish Population of the Bratsk Reservoir. *J. Ichthyol.* **2024**, *64*, 650–662. [CrossRef]
21. Louca, V.; Lindsay, S.W.; Lucas, M.C. Factors triggering floodplain fish emigration: Importance of fish density and food availability. *Ecol. Freshw. Fish* **2010**, *18*, 60–64. [CrossRef] [PubMed]
22. Li, K.; Tang, H.; Yuan, S. A field study of near-junction-apex flow at a large river confluence and its response to the effects of floodplain flow. *J. Hydrol.* **2022**, *610*, 127983. [CrossRef]
23. Xia, S.; Liu, Y.; Wang, Y.; Chen, B.; Jia, Y.; Liu, G.; Yu, X.; Wen, L. Wintering waterbirds in a large river floodplain: Hydrological connectivity is the key for reconciling development and conservation. *Sci. Total Environ.* **2016**, *573*, 645–660. [CrossRef] [PubMed]
24. Li, H.; Zhang, H.; Yu, L.; Cao, K.; Wang, D.; Duan, X.; Ding, F.; Mao, Z.; Wang, K.; Liu, S.; et al. Managing Water Level for Large Migratory Fish at the Poyang Lake Outlet: Implications Based on Habitat Suitability and Connectivity. *Water* **2022**, *14*, 2076. [CrossRef]
25. Qiang, Z.; Ya, F. Geometric properties of river cross sections and associated hydrodynamic implications in Wuhan-Jiujiang River reach, the Yangtze River. *J. Geogr. Sci.* **2009**, *19*, 58–66. [CrossRef]
26. Yang, Y.; Li, M.; Liu, W. Relationship between potential waterway depth improvement and evolution of the Jingjiang Reach of the Yangtze River in China. *J. Geogr. Sci.* **2023**, *33*, 547–575. [CrossRef]
27. Lanés, L.E.K.; Rolon, A.S.; Stenert, C. Effects of an artificial and annual opening of a natural sandbar on the fish community in a coastal lagoon system: A case study in Lagoa do Peixe floodplains, southern Brazil. *J. Appl. Ichthyol.* **2015**, *31*, 321–327. [CrossRef]
28. Chen, M.; Yu, D.; Lian, Y. Population abundance and habitat preference of the Yangtze finless porpoise in the highest density section of the Yangtze River. *Aquatic Conservation Marine and Freshwater Ecosystems* **2020**, *30*, 1088–1097. [CrossRef]
29. Fernandes, C.C. Amazonian ecology: Tributaries enhance the diversity of electric fishes. *Science* **2004**, *305*, 1960–1962. [CrossRef] [PubMed]
30. Röpke, C.P.; Amadio, S.A.; Winemiller, K.O. Seasonal dynamics of the fish assemblage in a floodplain lake at the confluence of the Negro and Amazon Rivers. *J. Fish Biol.* **2016**, *30*, 1088–1097. [CrossRef] [PubMed]
31. Gayoso, A.M.; Podesta, G.P. Surface hydrography and phytoplankton of the Brazil-Malvinas currents confluence. *J. Plankton Res.* **1996**, *18*, 941–951. [CrossRef]
32. Jiang, X.; Li, M.; Yang, S.; Lin, P.; Chang, T.; Wang, C.; Zhang, C.; Gao, X. Temporal variation of fish biodiversity in Poyang Lake. *Acta Hydrobiol. Sin.* **2023**, *47*, 376–388. [CrossRef]
33. Li, H.; Cao, K.; Wang, D.; Ding, F.; Mao, Z.; Yu, L.; Wang, K.; Duan, X.; Jia, C.; Chen, D.; et al. Habitat suitability analysis of fish community in the channel connecting the Poyang Lake and the Yangtze River in winter. *J. Fish. Sci.* **2022**, *29*, 341–354.

34. Tan, Z.; Li, Y.; Zhang, Q.; Liu, X.; Song, Y.; Xue, C.; Lu, J. Assessing effective hydrological connectivity for floodplains with a framework integrating habitat suitability and sediment suspension behavior. *Water Res.* **2021**, *201*, 117253. [CrossRef] [PubMed]
35. Zhu, Q. *Studies on the Changes of Fish Community Structure in Summer and Autumn in the Tongjiang Waterway of Poyang Lake and the Otolith and Growth of Young the Four Famous Domestic Fishes*; Nanchang University: Nanchang, China, 2011. [CrossRef]
36. Li, H.; Wang, K.; Yu, X.; Gao, L.; Jia, C.; Guo, J.; Duan, X.; Wang, S.; Chen, D.; Wang, D.; et al. Hydroacoustic surveys on spatial-temporal distribution of fishes during early fishing ban period in poyang Lake. *Acta Hydrobiol. Sin.* **2023**, *47*, 147–155. [CrossRef]
37. Jiang, X.; Li, L.; Yang, S. Study on the structural characteristics and temporal changes of fish assemblages in the Poyang Lake. *Resour. Environ. Yangtze River Basin* **2022**, *31*, 588–601. [CrossRef]
38. González-Sansón, G.; Aguilar-Betancourt, C.; Kosonoy-Aceves, D.; Lucano-Ramírez, G.; Ruiz-Ramírez, S.; Flores-Ortega, J.R.; Hinojosa-Larios, Á.; Silva-Bátiz, F.D.A. Species and size composition of fishes in Barra de Navidad lagoon, Mexican central Pacific. *Rev. De Biol. Trop.* **2014**, *62*, 142–157. [CrossRef]
39. Chen, W.; He, G.; Wu, B. Spatial distribution and biomass assessment of fish in the channel connecting the Lake Poyang and the Yangtze River. *J. Lake Sci.* **2017**, *29*, 923–931.
40. Shi, Y.; Yu, L.; Zhou, X.; Gao, L.; Zhu, F.; Yang, J.; Chen, D.; Wang, K.; Duan, X. Spatial-temporal distribution characteristics of fish resources in Yichang-Chenglingji River section of the Yangtze River in the early period of fishing ban. *Acta Hydrobiol. Sin.* **2024**, *48*, 546–557. [CrossRef]
41. Li, X. *Spatial and Temporal Distribution Characteristics of Fish in the Mainstream of the Upper Yangtze River Reserve Jiangan-Chongqing Section in the Early Period of Closed Fishing*; Southwest University: Chongqing, China, 2022; pp. 23–44.
42. Hu, F.; Zhu, T.; Gong, J. Acoustic study of fish resources in Xiangjiaba Reservoir in the lower reaches of Jinsha River during the early period of ten-year fishing ban. *J. Fish. China* **2023**, *47*, 233–244.
43. Jia, C. *Hydroacoustic Studies on Fishes in the Middle Yangtze from Yichang to the East Dongting Lake*; Ocean University Shanghai: Shanghai, China, 2022; p. 13.
44. Liu, F.; Wang, Z.; Xia, J. Changes in fish resources 5 years after implementation of the 10-year fishing ban in the Chishui River, the first river with a complete fishing ban in the Yangtze River Basin. *Ecol. Process.* **2023**, *12*, 51. [CrossRef]
45. Zhang, H.; Kang, M. Rapid change in Yangtze fisheries and its implications for global fresh water ecosystem management. *Fish Fish.* **2020**, *21*, 601–620. [CrossRef]
46. Wu, Z.; Zhang, H.; Zhang, G. Study on the growth characteristics and population parameters of *Siniperca chuatsi* before and after the fishing ban in Poyang Lake. *J. Jiangxi Agric. Univ.* **2024**, *46*, 220–231.

fishes

MDPI

Article

Evaluating the Effectiveness of an Acoustic Camera for Monitoring Three Large Jellyfish Species in the Coastal Waters of Liaodong Bay, China

Bin Wang †, Xiuze Liu †, Jing Dong *, Aiyong Wang , Chao Feng, Yanzhao Xu, Depu Zhang and Zhongfang Zhao

Key Laboratory of Protection and Utilization of Aquatic Germplasm Resource, Ministry of Agriculture and Rural Affairs, Liaoning Key Laboratory of Marine Biological Resources and Ecology, Dalian Key Laboratory of Conservation of Fishery Resources, Liaoning Ocean and Fisheries Science Research Institute, Dalian 116023, China; wangbin81321@163.com (B.W.); 13384119670@163.com (X.L.); wangaiyong@live.com (A.W.); fengch2021@163.com (C.F.); hkyxyzwow@163.com (Y.X.); zdp1769052195@163.com (D.Z.); shay.d.dorian@gmail.com (Z.Z.)

* Correspondence: dongj660228@163.com

† These authors contributed equally to this work.

Abstract: A survey was conducted to evaluate the effectiveness of adaptive resolution imaging sonar (ARIS), also known as an acoustic camera, for monitoring large jellyfish in the Liaodong Bay area, China. The abundance and vertical distribution of large jellyfish species, such as *Nemopilema nomurai*, *Aurelia coerulea*, and *Cyanea nozakii*, were obtained from acoustic camera observation images, and the effectiveness of the acoustic camera method was determined. The acoustic camera method provided visual information on the number of large jellyfish and their positions in the water column and demonstrated that they were more frequently located in the mid-upper water column of the surveyed area. The results show that it is possible to identify three different types of large jellyfish using acoustic camera sonar images, based on their size, shape, outline, and movement trajectory. The acoustic camera method enables the effective monitoring of jellyfish abundance and enables the observation of their vertical distribution, demonstrating its suitability for monitoring large jellyfish in shallow waters. The results show that observations through an acoustic camera can be used to study the horizontal and vertical spatial distribution characteristics of large jellyfish and to observe their behavior.

Keywords: adaptive resolution imaging sonar (ARIS); *Nemopilema nomurai*; *Aurelia coerulea*; *Cyanea nozakii*; abundance; distribution

Key Contribution: Images of different jellyfish species obtained through an acoustic camera were analyzed; and the results indicated that it was possible to distinguish three distinct types of large jellyfish in the images based on their size, shape, outline, and movement trajectory. We thus validated the effectiveness of using an acoustic camera to monitor large jellyfish populations in coastal waters.

1. Introduction

In recent years, the global decline in fisheries has been accompanied by a surge in jellyfish populations. Marine ecosystems have likely been altered by factors such as climate change, environmental pollution, overfishing, and marine engineering activities, resulting in favorable conditions for jellyfish proliferation [1–6]. Since the late 20th century, large-scale jellyfish blooms have been occurring increasingly frequently in many parts of the world, significantly impacting marine ecosystems, fisheries, coastal industries, and coastal tourism [7–12]. The coastal waters of China are among those most severely affected by large-scale jellyfish blooms [12–15]. However, accurately monitoring the abundance, biomass, and spatial distribution of jellyfish populations remains a crucial challenge in assessing ongoing jellyfish bloom occurrences. Internationally, monitoring jellyfish populations is a difficult task due to their high water content, large size, patchy distribution, and presence at varying depths in the water column. These factors have often led to inaccurate estimations in monitoring survey results [16].

Traditional survey methods for monitoring jellyfish populations primarily include net sampling [13–15,17] and visual observation [18–21], where the distribution and resource variation in jellyfish are studied, respectively. The advantages of these methods are that they provide a standardized sampling approach, are easy to operate, and visually reflect the spatial distribution characteristics of large jellyfish. However, they lack information and analytical data about swimming, aggregation, and the vertical movement patterns of jellyfish underwater. Hence, new technologies, such as underwater video [16] and aerial imaging [22] have been developed to monitor large jellyfish; however, both techniques have advantages and disadvantages [23]. The acoustic survey of fishery resources demonstrates advantages such as high efficiency, broad adaptability, and causing no harm to biological resources [24]. Internationally, acoustic technology is gradually being applied to the monitoring and surveying of large jellyfish, achieving positive results [25–29]. In the past two decades, acoustic cameras have been widely used in international fishery management, playing a crucial role in fish behavior research [30–35], fish population observation [36–41], fish length measurement [32–42], and even the survey of aquatic plants [43,44] and benthic animals [45] in shallow lakes. Furthermore, acoustic cameras have also been employed in the past for the monitoring and investigation of a specific type of large jellyfish, such as *Nemopilema nomurai* in offshore waters [46,47], or *Aurelia aurita* s.l. in a brackish water lake [48,49]. Given the limitations of traditional methods, and considering the successful application of acoustic cameras in various underwater observation studies, we hypothesize that acoustic cameras could offer a more effective alternative for monitoring jellyfish populations in Liaodong Bay by providing detailed visual information on jellyfish abundance and distribution that traditional methods cannot.

Throughout the past decade, Chinese scholars have gradually begun utilizing acoustic cameras in fish resource surveys and for behavioral observations in different water bodies [50–55], demonstrating the promising application of acoustic cameras in fishery management in China. In China, net-based surveys [13,14,56–58] have traditionally served as the primary means for monitoring large jellyfish. Subsequently, relevant reports on visual surveys [59] of large jellyfish have been published. In recent years, low-altitude Unmanned Aerial Vehicle remote sensing [60] and acoustic investigation [61–64] have gradually been employed for the monitoring and investigation of large jellyfish in China. Large jellyfish blooms have occurred frequently in the last two decades in Liaodong Bay, Bohai Sea, China. The dominant species found in Liaodong Bay were *N. nomurai*, *Aurelia coerulea*, and *Cyanea nozakii* [65]. However, since the blooms of *C. nozakii* in 2004 [66], economically valuable jellyfish (*Rhopilema esculentum*) yields in the region have declined significantly. From 2004 to the present, we conducted annual monitoring surveys of large jellyfish in Liaodong Bay using anchor drift nets, focusing on species such as *R. esculentum*, *N. nomurai*, *A. coerulea*, and *C. nozakii*, and achieved good monitoring results [67–70]. However, it is difficult to conduct intuitive and real-time monitoring of the water layer distribution and diurnal distribution patterns in large jellyfish. This study aims to evaluate the effectiveness of acoustic cameras, specifically adaptive resolution imaging sonar (ARIS), in overcoming these challenges by accurately monitoring the abundance, biomass, and spatial distribution of three large jellyfish species in Liaodong Bay. Since 2017, we have explored using an acoustic camera to monitor the giant jellyfish *N. nomurai* [71]. In the current study, we simultaneously used an acoustic camera and anchor drift nets to monitor the abundance and distribution of various large jellyfish species in Liaodong Bay. In this study, images of various large jellyfish species were obtained using an acoustic camera, and the advantages and disadvantages of the acoustic camera method over anchor drift nets with regard to large jellyfish monitoring were ascertained. In addition, we validated the effectiveness of an acoustic camera for monitoring various large jellyfish species in coastal waters.

2. Materials and Methods

2.1. Study Area and Sampling Methods

Based on the results of previous surveys that identified the coastal waters near Yingkou in Liaodong Bay as an area where large jellyfish frequently appear, we selected this region as the monitoring area to conduct surveys on 3 July and 9 August 2018. The maximum water depth in the survey area was 15 m. In accordance with the results of an investigation of large jellyfish using anchor drift nets, carried out a few days prior to this survey, three regions with relatively high quantities of large jellyfish were chosen as the locations for the net gear survey and sonar observation, using the fishing vessel Liao Ying Fishery 15228 (power 88.2 kW). Following the local standard of Liaoning Province, "Technical specification for monitoring survey and biomass assessment of disaster-causing jellyfish in offshore areas" (DB21/T 2823-2017) and the "Technical specification for monitoring survey and biomass assessment of disaster-causing jellyfish in offshore areas" (DB21/T 3368-2021) [72,73], traditional net sampling was conducted to monitor the presence of large jellyfish. In July, the sampling gear used was a *Sardinella zunasi* anchor drift net (60 m × 8 m; mesh size: 3 cm), and in August, a jellyfish anchor drift net (60 m × 10 m; mesh size: 10 cm) was used. Both types of nets are anchor drift nets that use sea currents to capture large jellyfish. Three nets were connected end-to-end for deployment at each location, and after 1 h, the nets were retrieved to count the number of large jellyfish captured. The average number of large jellyfish caught in the three nets was then calculated. In addition, after the deployment of the anchor drift nets, an acoustic camera observation transect was established approximately 50 m from the nets in the direction of their deployment. The amount of large jellyfish in the net was calculated using the catch number and the filtration volume for each operation. The filtration volume was calculated using the net height, net width, current velocity, soak time, and the rate of net height deformation, which was assumed to be around 50%, according to the experience of the local fishers [58]. The observation transect was located on the side where the jellyfish entered the net walls (see Figure 1), and each transect was monitored for 1 h. The water current velocity was recorded from the side of the vessel for 3 min using an electromagnetic current direction meter (AEM213-D, JFE Advantech Co., Ltd., Nishinomiya, Japan) in the middle layer of the water during net deployment.

2.2. Data Acquisition

Adaptive resolution imaging sonar (Aris Explorer 1800, Sound Metrics Corp, Bellevue, WA, USA) was utilized in identification mode, a high-frequency mode of 1.8 MHz. The effective depth of the observations corresponded to the water depth. The sonar beam had a horizontal width of 0.3°, a vertical width of 14°, an overall field of view of 29°, and a total of 96 beams when operated in the high-frequency mode. The sonar images (receiver gain: 15 dB) were transferred from the unit to a computer via control software at 7.3–15 frames s^{-1} and saved on a hard drive for later review. A custom-made bracket was used to fix the main acoustic camera unit to the side of the vessel. The angle was adjusted such that the main unit (sensor) was submerged below the water surface and the probe was located at a depth of 0.3 m below the water surface, where it remained perpendicular to the surface (see Figure 2). Given that waves could influence the results of sonar observations, we elected to carry out sonar observations in weather conditions of near calmness on the sea surface (with wind speeds lower than Level 2 and wave heights less than 0.2 m) to minimize the impact of waves. After deploying the nets, the survey vessel proceeded along the designed transect at a speed of 0.7–0.9 m/s (2.5–3.5 km/h). Acoustic camera data were collected for 1 h at each of the three transects. The number of large jellyfish, their distribution in the water column, and the corresponding times were manually recorded from videos taken using acoustic cameras. The number of large jellyfish observed in these acoustic camera videos was counted at 1 min intervals to calculate their abundance.

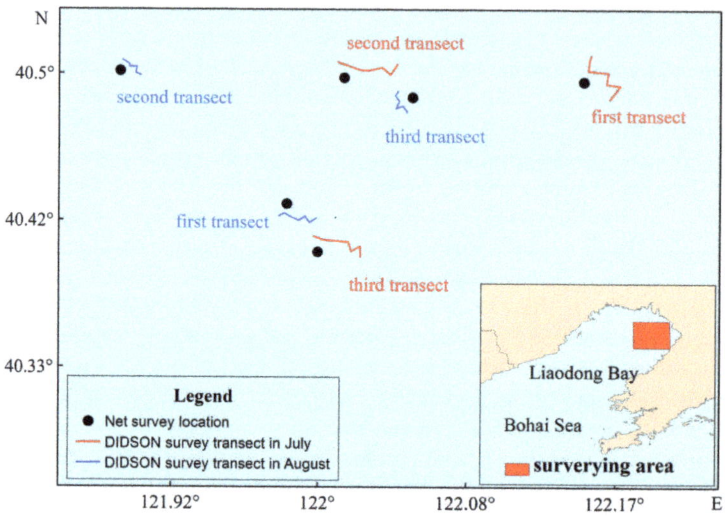

Figure 1. Map showing the location of the transects of three large jellyfish species monitored using acoustic cameras and nets in the coastal waters of Liaodong Bay, China, in 2018.

The manual counting of large jellyfish during sonar observations was conducted using the ARIScope software (Version 2.5) that comes with the adaptive resolution imaging sonar. As the morphological characteristics of the umbrella, oral arms, and tentacles of the three species of large jellyfish are distinctly different, we can distinguish them by their individual size, shape, outline, and movement trajectory in the acoustic camera images (see Figure 3); when the objects were difficult to identify, we also checked several frames both before and after the analytical frame to confirm the identification based on bell pulsation and movement trajectory. The bell diameter of the large jellyfish was measured by drawing a horizontal box using the right-click selection tool in the ARIScope software during manual counting. The horizontal box was positioned such that its diagonal aligned with the two lower edges of the jellyfish's umbrella, and the length of this diagonal corresponded to the umbrella diameter. In this study, we examined the spatial distribution as well as the population dynamics over two months of *N. nomurai* (average bell diameters of 33.65 ± 9.40 cm in July and 50.04 ± 12.08 cm in August), *A. coerulea* (average bell diameters of 15.67 ± 2.98 cm in July and 18.83 ± 4.08 cm in August), and *C. nozakii* (average bell diameters of 23.44 ± 5.78 cm in July and 39.68 ± 10.44 cm in August), which were measured using ARIScope software. All the large jellyfish obtained via the anchor drift net were identified by species, and their numbers were counted, which were taken as the baseline data for the sonar observations of the abundance of the different large jellyfish. The detailed elaboration of the morphological characteristics of the three large jellyfish types in the acoustic camera images is introduced in Section 3.1 of the Results.

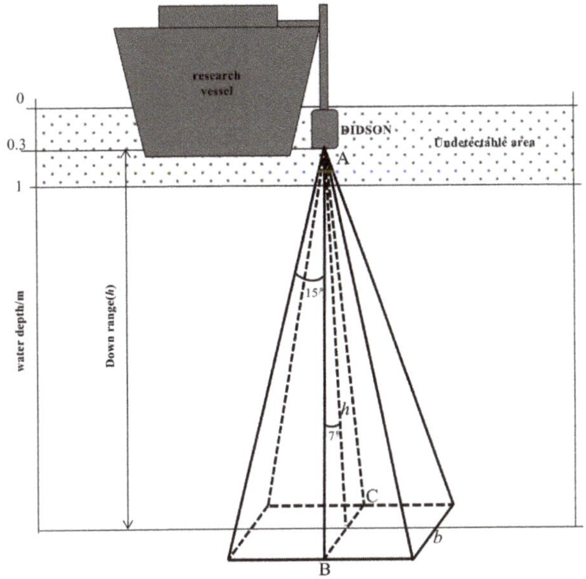

Figure 2. Schematic representation of the pyramid-shaped acoustic camera view field.

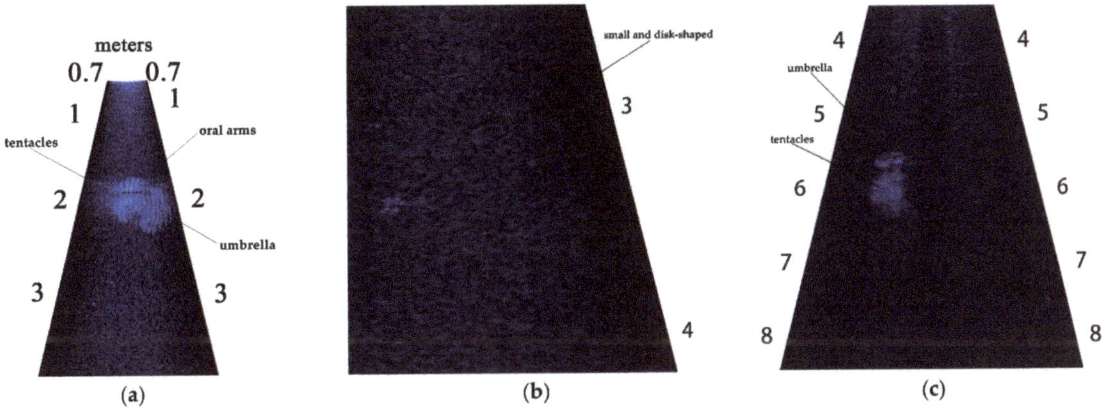

Figure 3. The enlarged acoustic camera images of the large jellyfish *N. nomurai*, *A. coerulea*, and *C. nozakii* in July and August in the coastal waters of Liaodong Bay. (**a**) The umbrella, tentacles, and oral arms of *N. nomurai* in acoustic camera images; (**b**) *A. coerulea* is seen as small and disk-shaped in acoustic camera images; (**c**) the umbrella and tentacles of *C. nozakii* in acoustic camera images. The white shade in the image delineates the outline of a large jellyfish.

2.3. Data Processing

The abundance of large jellyfish monitored using an acoustic camera or nets was calculated by dividing the number of observed or captured jellyfish by the survey volume using the following formula:

$$\rho = 1000\, n/V \tag{1}$$

where ρ is the abundance of large jellyfish (ind/1000 m^3); n is the number of large jellyfish detected with the sonar or captured with the net (ind); and V is the survey volume of the sonar or nets (m^3).

The survey volume for the acoustic cameras was calculated by multiplying the area scanned using the camera during its operation by the transect length. We referred to the method used by Mo et al. [74] during acoustic surveys of fisheries conducted in the Dongqing and Guangzhao Reservoir areas of the Beipan River, and the water bodies detected in each acoustic camera frame were approximated as a tetrahedron (Figure 2). The sonar survey volume was calculated using the following formulas:

$$V_{\text{acoustic camera}} = S_{\text{ABC}}\, vt, \tag{2}$$

$$S_{\text{ABC}} = 0.5\, bh, \tag{3}$$

$$b = 2h \tan 7^\circ, \tag{4}$$

where $V_{\text{acoustic camera}}$ is the sonar-detected water volume (m^3); S_{ABC} is the area of the central vertical cross-section of the tetrahedron in the vertical direction of travel (m^2); v is the vessel speed during sonar scanning (m/s); t is the sonar scanning time (s); b is the width of the farthest cross-section detected by the sonar (m); and h is the maximum imaging distance, representing the actual effective depth scanned by the sonar (m) (Figure 4a).

The net survey volume was calculated by multiplying the cross-sectional area of the net during the survey by the distance through which the current flowed. The net survey volume was calculated using the following formula:

$$V_{\text{net}} = LH\, v_{\text{h}}\, T, \tag{5}$$

where V_{net} is the survey volume of the anchor drift net (m^3); L is the width of the net spread in the sea (m), which was 50 m in this study; H is the height of the net in the sea (m), the rate of net height deformation, which was assumed to be around 50% when fully extended under the impact of ocean currents, according to the experience of the local fishers [58]; v_{h} is the flow velocity through the net (m/s), which had an average speed of 0.265 m/s based on field measurements in this study; and T is the net survey time (s) (Figure 4b).

The correlation between the abundance of large jellyfish observed through the acoustic camera and that surveyed using the anchor drift net was analyzed using Pearson's correlation analysis via the statistical software SPSS 22 to evaluate the effectiveness of the acoustic method.

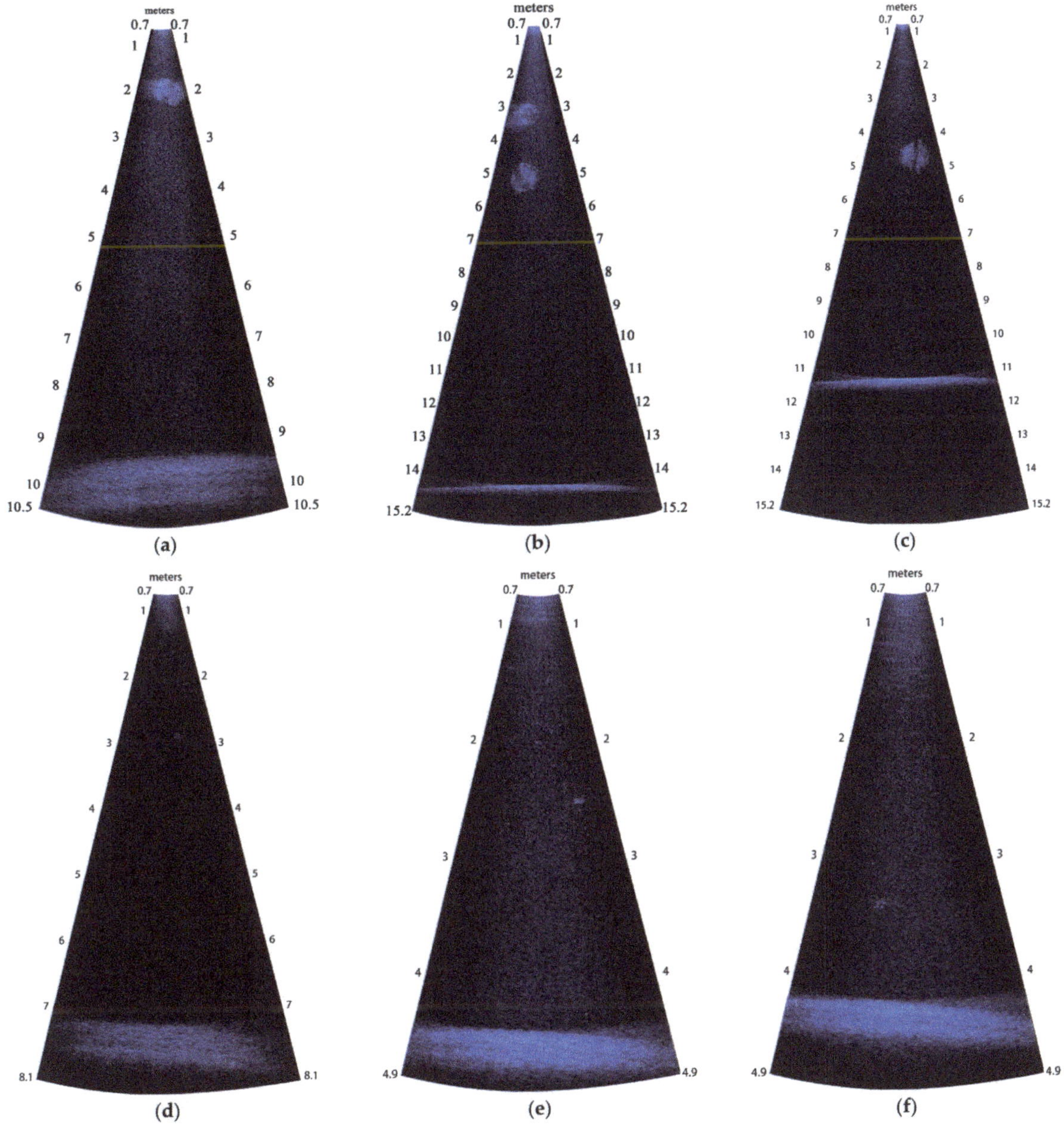

Figure 4. *Cont.*

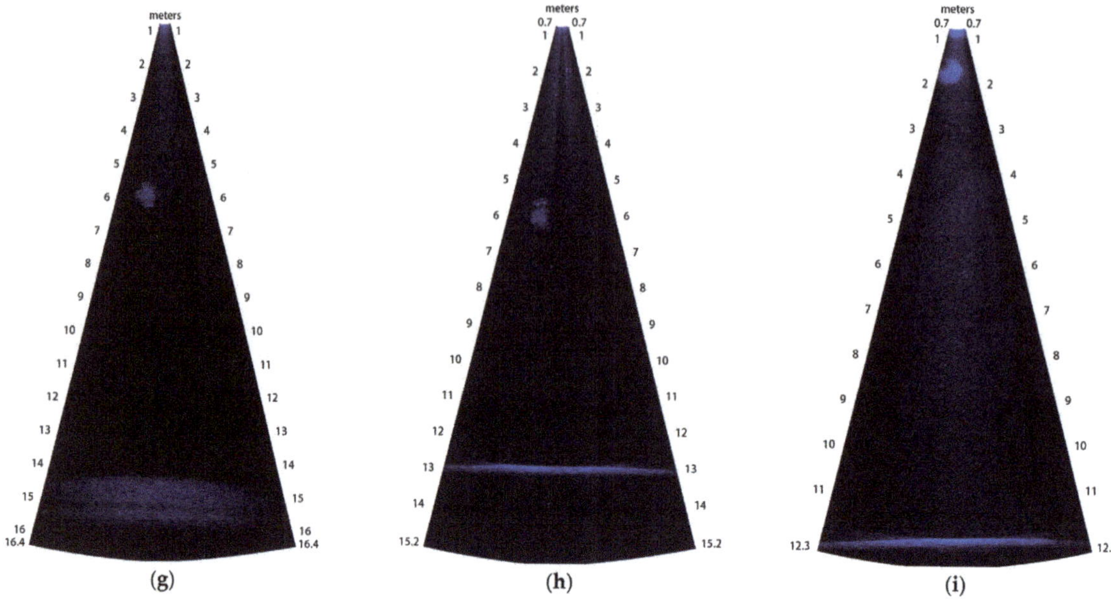

Figure 4. Acoustic camera images of the large jellyfish species *N. nomurai*, *A. coerulea*, and *C. nozakii* in July and August in coastal waters of Liaodong Bay. (**a–c**) Acoustic camera images of *N. nomurai*; (**d–f**) Acoustic camera images of *A. coerulea*. (**g–i**) Acoustic camera images of *C. nozakii*. The white shade in the image delineates the outline of a large jellyfish, while the white horizontal line at the bottom of the figure represents the seabed.

3. Results

3.1. Morphological Characteristics of the Three Large Jellyfish Species in Acoustic Camera Images

Figure 5 shows nine images extracted from the acoustic camera videos. The top three images, middle three images, and bottom three images correspond to observations of *N. nomurai*, *A. coerulea*, and *C. nozakii*, respectively.

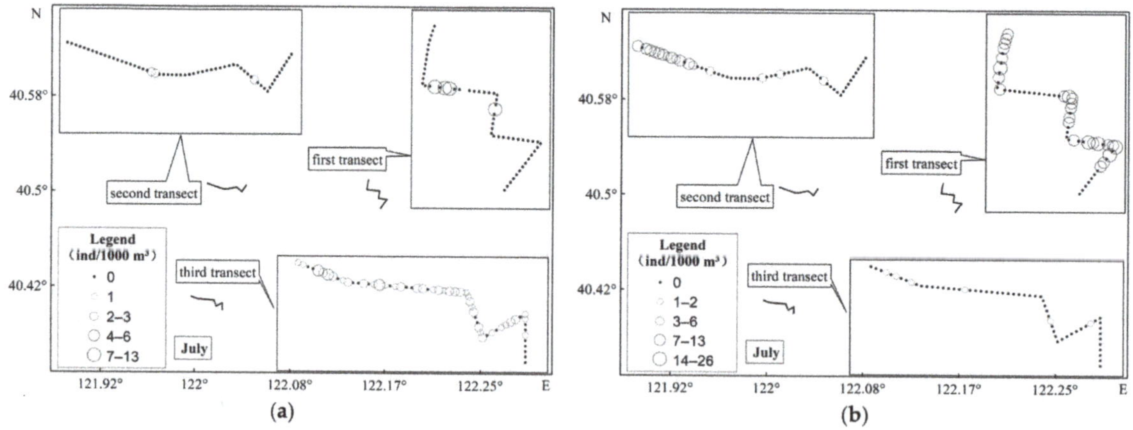

Figure 5. *Cont.*

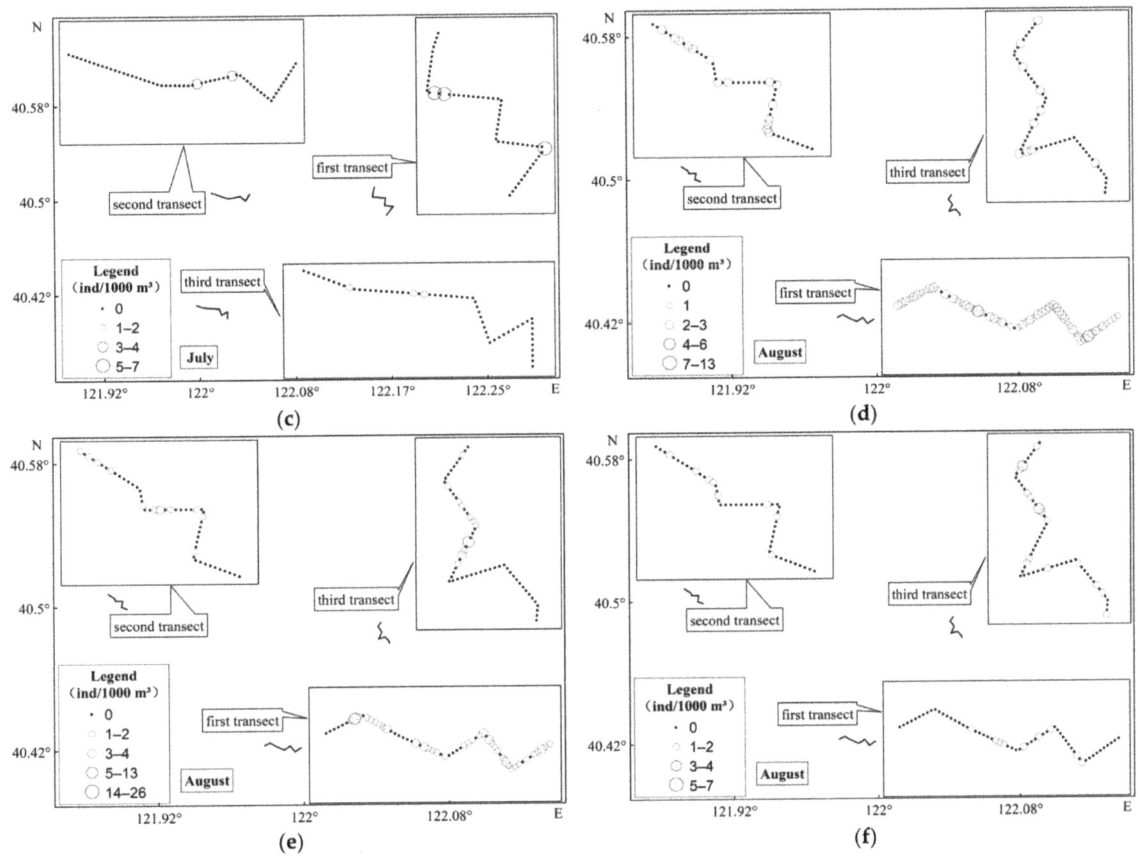

Figure 5. Horizontal distribution of *N. nomurai*, *A. coerulea*, and *C. nozakii* medusae abundance along three transects, obtained through acoustic camera images in July and August in the coastal waters of Liaodong Bay. (**a**) The horizontal distribution of *N. nomurai* in July. (**b**) The horizontal distribution of *A. coerulea* in July. (**c**) The horizontal distribution of *C. nozakii* in July. (**d**) The horizontal distribution of *N. nomurai* in August. (**e**) The horizontal distribution of *A. coerulea* in August. (**f**) The horizontal distribution of *C. nozakii* in August.

Nemopilema nomurai belongs to the Rhizostomae order and is the largest jellyfish species in the East Asian waters. As shown in Figure 4a–c, *N. nomurai* individuals are relatively large, and the acoustic camera sonar images clearly reveal the outline of the jellyfish. The arc-shaped contour of the umbrella and the oral arms and tentacles of *N. nomurai* are distinct. Therefore, *N. nomurai* is the most easily identifiable of the three types of jellyfish.

Aurelia coerulea belongs to the Semaeostomeae order. The bell diameter of *A. coerulea* was significantly smaller than that of *N. nomurai* and *C. nozakii*, and the images captured by the sonar were less clear. In images acquired from the acoustic camera video, *A. coerulea* often appears as a small disc or oval shape, and the oral arms were shorter and hardly visible in the acoustic camera images (Figure 4d–f).

Cyanea nozakii belongs to the Semaeostomeae order, and the umbrella of *C. nozakii* is flatter with longer tentacles of similar width. *Cyanea nozakii* individuals were also relatively large with an average bell diameter smaller than that of *N. nomurai*. As shown in Figure 4d,h,i, the shapes and outlines of *C. nozakii* individuals were distinct from those of

N. nomurai. They had a cotton ball-like shape in the sonar video, which helped distinguish them from *N. nomurai*.

Therefore, combined with the above morphological features, we were able to identify the three large jellyfish species in acoustic camera images based on their size, shape, outline, and movement trajectory.

3.2. Horizontal Distribution of Large Jellyfish Abundance

The abundance of the three large jellyfish species observed using an acoustic camera over a period of 1 h across the three survey transects showed different trends in July and August. The left side of the survey transect was the front section, and the right side was the back section. The horizontal distribution characteristics of the three survey transects of the three large jellyfish species varied as follows.

According to the acoustic camera survey conducted in July, *N. nomurai* were scattered in the middle sections of the first survey transect and the second survey transect but were highly abundant and relatively uniformly distributed in the third survey transect. *Aurelia coerulea* were mainly distributed in the middle and front sections of the first survey transect and the front section of the second survey transect but were scattered throughout the third survey transect. *Cyanea nozakii* had a scattered distribution throughout all three survey transects (see Figure 5a–c).

According to the acoustic camera survey conducted in August, *N. nomurai* was present only with a scattered distribution in part of the third survey transect and in the front and middle sections of the second survey transect. *N. nomurai* was highly abundant in the first survey transect and was distributed throughout almost every part of the transect. *Aurelia coerulea* was sparsely distributed throughout parts of the second survey transect and the third survey transect and was more abundant in the first survey transect. *Cyanea nozakii* was distributed sporadically across the three survey transects (see Figure 5d–f).

3.3. Study of the Vertical Distribution of Large Jellyfish

Sonar images showed the following vertical distributions of *N. nomurai* in July: along Transect 1, up to a depth of 5 m, *N. nomurai* mainly inhabited the middle water layer (2.1–4.0 m); in Transect 2, up to a depth of 8 m, they primarily resided in the upper to middle water layers (2.1–6.0 m); and in Transect 3, up to a depth of 14 m, they were more abundant in the upper to middle water layers (1.1–6.0 m) (see Figure 6a). Overall, in July, there were more *N. nomurai* present in the upper to middle water layers. The following was observed in August: in Transect 1, up to a depth of 15 m, the *N. nomurai* individuals were identified in almost all the water layers (1.1–14.0 m) and were most abundant in the middle water layers (8.1–11.0 m); in Transect 2, up to a depth of 14 m, *N. nomurai* inhabited the upper to middle water layers (0.7–7.0 m); and in Transect 3, up to a depth of 10 m, they were more abundant in the upper to middle water layers (0.7–6.0 m). In August, *N. nomurai* was still more abundant in the upper and middle water layers (see Figure 6b). Overall, in both July and August, *N. nomurai* were most commonly found in the upper and middle water layers of the surveyed area, whereas none were found in the bottom water layers.

Sonar images showed the following vertical distribution of *A. coerulea* in July: in Transect 1, up to a depth of 5 m, *A. coerulea* mainly inhabited the middle water layer (1.1–4.0 m); in Transect 2, up to a depth of 8 m, they primarily resided in the upper to middle water layers (1.1–5.0 m); and in Transect 3, up to a depth of 14 m, they were more abundant in the upper to middle water layers (4.1–7.0 m). Overall, in July, there were more *A. coerulea* in the upper to middle water layers (see Figure 6c). The following was observed in August: in Transect 1, up to a depth of 15 m, most of the *A. coerulea* inhabited the upper and middle water layers (3.1–11.0 m); in Transect 2, up to a depth of 14 m, *A. coerulea* were found across almost all the water layers (0.7–12.0 m); in Transect 3, up to a depth of 10 m, *A. coerulea* mainly inhabited the upper to middle water layer (0.7–8.0 m). In August, *A. coerulea* was still more abundant in the upper and middle water layers (see Figure 6d). Overall, in both July and August, *A. coerulea* were more commonly found in the upper and middle water layers of the surveyed area, whereas none were found in the bottom water layers.

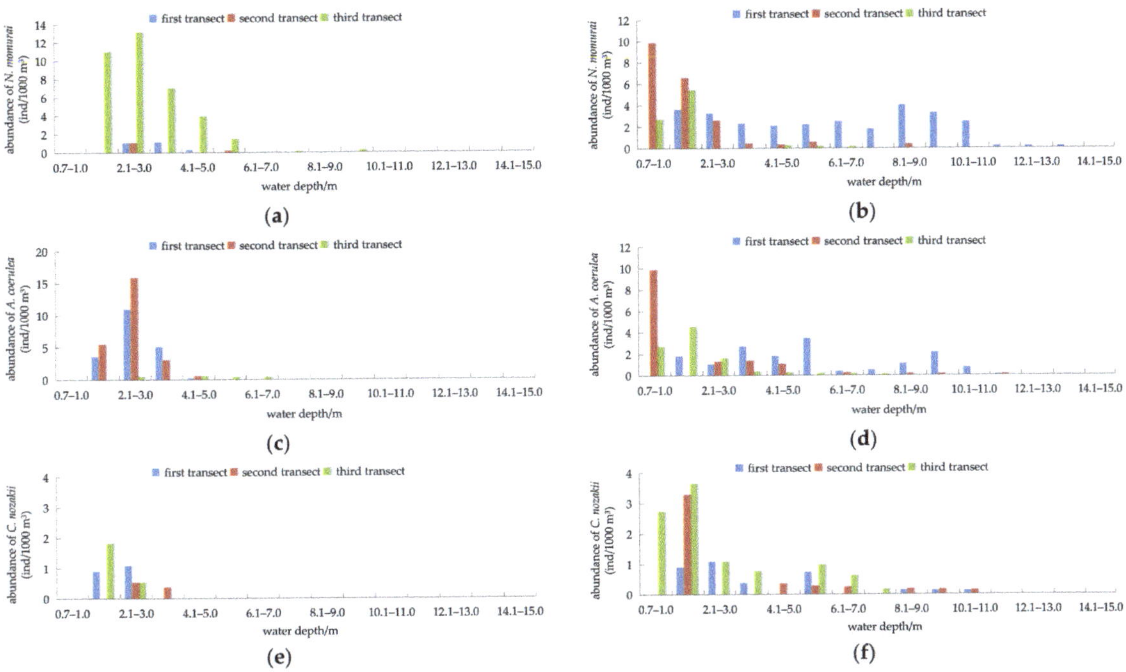

Figure 6. The vertical distribution of the large jellyfish *N. nomurai, A. coerulea,* and *C. nozakii* medusae obtained through acoustic camera images in July in Liaodong Bay. (**a**) *N. nomurai* in July; (**b**) *N. nomurai* in August; (**c**) *A. coerulea* in July; (**d**) *A. coerulea* in August; (**e**) *C. nozakii* in July; (**f**) *C. nozakii* in August.

Sonar images showed the following vertical distribution of *C. nozakii* medusae in July: in Transect 1, up to a depth of 5 m, *C. nozakii* inhabited the upper to middle water layers (1.1–3.0 m); in Transect 2, up to a depth of 8 m, they inhabited the upper to middle water layers; and in Transect 3, up to a depth of 14 m, they were more abundant in the upper water layers (1.1–3.0 m). Overall, in July, there were more *C. nozakii* in the upper to middle water layers (see Figure 6e). The following was observed in August: in Transect 1, up to a depth of 15 m, most of the *C. nozakii* inhabited the upper and middle water layers (1.0–11.0 m); in Transect 2, up to a depth of 14 m, *C. nozakii* were found in almost all of the water layers studied (1.1–11.0 m); and in Transect 3, up to a depth of 10 m, they were more abundant in the upper and middle water layers (0.7–8.0 m). In August, *C. nozakii* was still more abundant in the upper and middle water layers (see Figure 6f). Overall, in both July and August, *C. nozakii* was more commonly found in the upper to middle water layers of the surveyed area, whereas none were found in the bottom water layers.

3.4. Comparison of Jellyfish Abundances Obtained via Acoustic Camera Observations and Net Surveys

Based on the experience of local fishers, the deformation rate of the anchor drift net was assumed to be approximately 50% when fully extended under the impact of ocean currents [58], with the anchor drift nets usually placed in the middle water layers. To validate the accuracy and reliability of the acoustic camera data for large jellyfish, we compared the data from the anchor drift nets and acoustic cameras in the corresponding water layers. In July, the water depths in the three survey sections were 5, 8, and 14 m. Under the influence of ocean currents, the *Sardinella zunasi* anchor drift net deformed to a height of 4 m. We analyzed and compared the acoustic camera data for large jellyfish in the corresponding water layers of 1–5, 2–6, and 4–8 m in the three survey transects. In August, the water depths at the three survey transects were 10, 14, and 15 m. The jellyfish anchor drift net was deformed to a height of 5 m under the influence of ocean currents. We analyzed and compared the acoustic camera data for large jellyfish in the corresponding water layers of 2–7, 4–9, and 5–10 m across the three survey transects.

In July, 0.33, 0.5, and 47 *N. nomurai*, were captured using anchor drift nets over a period of 1 h near the three surveyed transects, respectively. The abundances recorded via the net surveys were 0.0017 ind/1000 m^3, 0.0026 ind/1000 m^3, and 0.25 ind/1000 m^3, with an average abundance of 0.08 $\pm$ 0.12 ind/1000 m^3. In August, 63, 8.33, and 14.67 *N. nomurai* were captured by the anchor drift nets over a period of 1 h along the three transects, respectively. The abundances recorded via net survey were 0.26 ind/1000 m^3, 0.03 ind/1000 m^3, and 0.06 ind/1000 m^3, with an average abundance of 0.12 $\pm$0.10 ind/1000 m^3. In July, 6, 3, and 20 *N. nomurai* were observed using sonar over a period of 1 h in the corresponding water layers of the three survey transects, respectively. The abundances recorded using sonar were 0.69 ind/1000 m^3, 0.26 ind/1000 m^3, and 1.15 ind/1000 m^3, with an average abundance of 0.70 $\pm$ 0.36 ind/1000 m^3. The average abundance of *N. nomurai* observed using sonar in July was 8.3 times higher than that recorded via the net surveys (Figure 7a). In August, 79, 5, and 3 *N. nomurai* were observed using sonar over a period of 1 h along the three surveyed transects, respectively. The abundances recorded using sonar were 2.89 ind/1000 m^3, 0.25 ind/1000 m^3, and 0.18 ind/1000 m^3, with an average abundance of 1.11 $\pm$ 1.26 ind/1000 m^3. The average abundance of *N. nomurai* monitored using sonar in August was 9.2 times higher than that recorded via the net surveys (Figure 7a). The abundance of *N. nomurai* observed using sonar in July was slightly lower than that observed in August. The net survey results were similar to those observed using the acoustic cameras, which also showed that the abundance of *N. nomurai* in July was lower than that in August, and that the abundance value obtained through the net survey was lower than that observed using the acoustic cameras. There was a significant correlation ($p < 0.05$) between the abundance of *N. nomurai* observed using the acoustic camera and that observed using net surveys in July and August.

In July, 140, 131, and 26 *A. coerulea* were captured by the anchor drift nets over a 1 h period near the three survey transects, respectively. The abundances recorded via the net surveys were 0.74 ind/1000 m^3, 0.69 ind/1000 m^3, and 0.14 ind/1000 m^3, with an average abundance of 0.52 $\pm$ 0.27 ind/1000 m^3. In August, 79, 4.67, and 4 *A. coerulea* were captured by the anchor drift nets over a period of 1 h in the three transects. The abundances recorded via the net surveys were 0.33 ind/1000 m^3, 0.02 ind/1000 m^3, and 0.02 ind/1000 m^3, with an average abundance of 0.12 $\pm$ 0.15 ind/1000 m^3. In July, 38, 39, and 6 *A. coerulea* were observed using sonar over a 1 h period across the three survey transects, respectively. The abundances recorded using sonar were 4.35 ind/1000 m^3, 3.35 ind/1000 m^3, and 0.34 ind/1000 m^3, with an average abundance of 2.68 $\pm$ 1.70 ind/1000 m^3. The average abundance of *A. coerulea* observed using sonar in July was 5.2 times higher than that recorded via the net surveys (Figure 7b). In August, 41, 5, and 7 *A. coerulea* were observed using sonar over a period of 1 h in the three survey transects, respectively. The abundances recorded using sonar were 1.50 ind/1000 m^3, 0.25 ind/1000 m^3, and 0.43 ind/1000 m^3, with an average abundance of 0.73 ± 0.55 ind/1000 m^3. The average abundance of *A. coerulea* monitored using sonar in August was 5.9 times higher than that recorded via the net surveys (Figure 7b). There was an extremely significant correlation ($p < 0.01$) between the abundance of *A. coerulea* observed using acoustic cameras and that observed via net surveys in July and August.

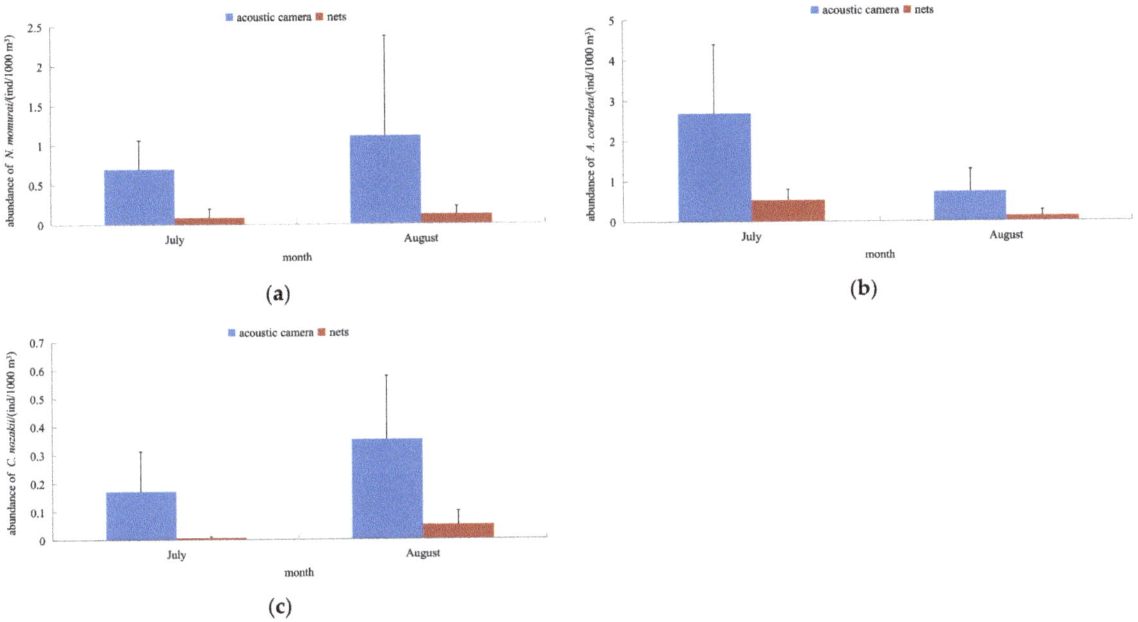

Figure 7. Comparison of the abundance of the three species of large jellyfish medusae through acoustic camera monitoring and net surveys in July and August in the coastal waters of Liaodong Bay. (**a**) *N. nomurai*; (**b**) *A. coerulea*; (**c**) *C. nozakii*.

In July, 2, 0, and 2 *C. nozakii* were captured by the anchor drift nets over a 1 h period near the three survey transects, respectively. The abundances recorded via net surveys were 0.0105 ind/1000 m^3, 0 ind/1000 m^3, and 0.0105 ind/1000 m^3, with an average abundance of 0.0070 ± 0.0049 ind/1000 m^3. In August, 5, 3.33, and 28.67 *C. nozakii* were captured by the anchor drift nets over a period of 1 h across the three transects, respectively. The abundances recorded via the net surveys were 0.02 ind/1000 m^3, 0.01 ind/1000 m^3, and 0.12 ind/1000 m^3, with an average abundance of 0.05 ± 0.05 ind/1000 m^3. In July, 3, 2, and 0 *C. nozakii* were observed using sonar over a period of 1 h along the three survey transects, respectively. The abundances recorded using sonar were 0.34 ind/1000 m^3, 0.17 ind/1000 m^3, and 0 ind/1000 m^3, with an average abundance of 0.17 ± 0.14 ind/1000 m^3. The average abundance of *C. nozakii* observed using sonar in July was 24.6 times higher than that recorded via the net surveys (Figure 7c). In August, 5, 4, and 11 *C. nozakii* were observed using sonar over a 1 h period in the three survey transects, respectively. The abundances recorded using sonar were 0.18 ind/1000 m^3, 0.20 ind/1000 m^3, and 0.67 ind/1000 m^3, with an average abundance of 0.35 ± 0.23 ind/1000 m^3. The average abundance of *C. nozakii* monitored using sonar in August was 6.8 times higher than that recorded using the net surveys (Figure 7c). There was a significant correlation ($p < 0.05$) between the abundance of C. *nozakii* observed using acoustic cameras and that observed using net surveys in July and August.

3.5. Bell Diameters of Large Jellyfish Observed via Acoustic Camera

The bell diameter of the large jellyfish was measured by drawing a horizontal box using the right-click selection tool in the ARIScope software during manual counting. The horizontal box was positioned such that its diagonal aligned with the two lower edges of the jellyfish's umbrella, and the length of this diagonal corresponded to the umbrella diameter. In July, the bell diameters of *N. nomurai* ranged from 15 cm to 55 cm, with an average of 33.65 ± 9.40 cm, and in August, they ranged from 16 cm to 74 cm, with an average of 50.04 ± 12.08 cm. Overall, in both July and August, *N. nomurai* bell diameters ranging from 15 to 74 cm were observed (Figure 8a).

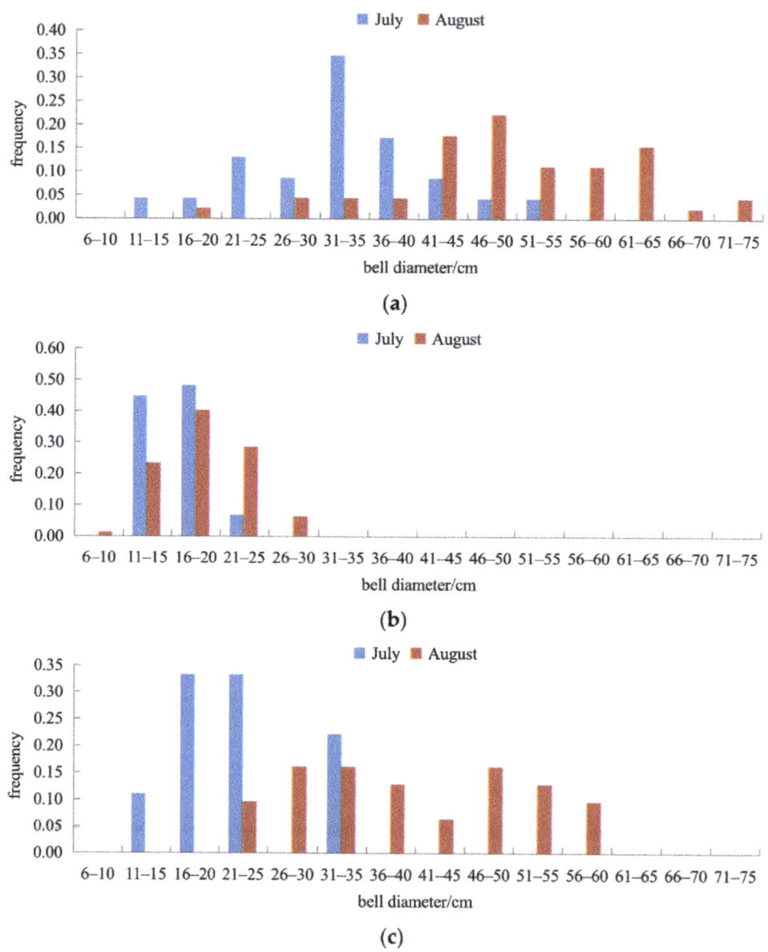

Figure 8. The bell diameter frequency distribution of the three large medusae jellyfish species in Liaodong Bay from the net samples in July and August in the coastal waters of Liaodong Bay. (**a**) *N. nomurai*; (**b**) *A. coerulea*; (**c**) *C. nozakii*.

In July, the bell diameters of *A. coerulea* ranged from 10 cm to 22 cm, with an average of 15.67 ± 2.98 cm, and in August, they ranged from 10 cm to 26 cm, with an average of

18.83 ± 4.08 cm. Overall, in both July and August, *A. coerulea* bell diameters ranging from 10 to 26 cm were observed (Figure 8b).

In July, the bell diameters of *C. nozakii* ranged from 15 cm to 33 cm, with an average of 23.44 ± 5.78 cm, and in August, they ranged from 22 cm to 56 cm, with an average of 39.68 ± 10.44 cm. Overall, in both July and August, *C. nozakii* bell diameters ranging from 15 cm to 56 cm were observed (Figure 8c).

4. Discussion

4.1. Analysis of the Effect of Applying an Acoustic Camera in Large Jellyfish Monitoring

Han and Uye [48] used acoustic cameras to monitor *Aurelia aurita* s.l. in a Japanese brackish coastal lake and observed bell diameters ranging from 4.1 cm to 19.6 cm (average of 13.1 cm). In comparison, the jellyfish observed in this study had larger bell diameters, making it easy to identify them in sonar images, which revealed a good observation effect. The use of acoustic cameras to monitor large jellyfish was effective owing to their ultra-high acoustic resolution, which allowed the outline, orientation, gonads, and appendages of large jellyfish to be clearly identified [48]. This makes these cameras an effective method for monitoring large jellyfish in shallow waters and for identifying their distribution in the water column. The findings of this study indicate that abundance assessments can be conducted using acoustic cameras without harming marine organisms, and that this technology clearly visualizes the water column distribution and number of large jellyfish, highlighting its excellent monitoring capabilities.

In this study, acoustic cameras were operated using a high-frequency identification mode at 1.8 MHz. In the acoustic camera images, the three different large jellyfish species were identified based on their size, morphology, outline, and movement trajectory. The sonar images clearly show the outlines of *N. nomurai* individuals, including their oral arms and tentacles. *C. nozakii* individuals are also large but can be distinguished from *N. nomurai* based on their shapes, outlines, and movement trajectories. Although *A. coerulea* individuals were smaller than those of the other two species, their individual size, outline, and movement trajectory could be distinguished using the acoustic camera. However, the images of *A. coerulea* were not as clear as those of *N. nomurai* and *C. nozakii*. In particular, it was easy to confuse the smaller *A. coerulea* with other debris in the acoustic camera images.

Makabe et al. [49] found that the smallest bell diameter of *Aurelia aurita* s.l. identified using an acoustic camera was 5 cm. In their study, Baumgartner et al. [32] reported that the acoustic camera's detection performance was poor for fish smaller than 7.5 cm, demonstrating the inefficiency of acoustic cameras in observing smaller fish species. Jůza et al. [75] found that fish less than 5 cm in length could not be clearly detected using acoustic cameras, resulting in a severe underestimation of their abundance. Smith et al. [76] set a detection threshold of 4 cm and noted that most fish at their study site were very small (<10 cm), potentially leading to an underestimation of the small juvenile fish community in their study. Therefore, acoustic cameras have a minimum detectable target size. In this study, some images of smaller individuals of large jellyfish species were unclear. This is related to the weak acoustic scattering intensity of the jellyfish [77]. In addition, species identification errors may occur in smaller individuals. Large jellyfish have different body shapes in the sea, and acoustic cameras only capture the dorsal side of the jellyfish bell at times, thus failing to effectively display the oral arms, tentacles, and other features. Although *N. nomurai*, *A. coerulea*, and *C. nozakii* can be differentiated based on their sizes, shapes, outlines, and movement trajectories, some jellyfish that are similar in size and shape to *R. esculentum* can sometimes be difficult to differentiate, especially when the individuals are smaller. As the current study also used an anchor drift net survey and found no *R.*

esculentum in the surveyed area, all individuals were considered *N. nomurai*, and were subsequently identified as such.

4.2. Advantages and Disadvantages of Using Acoustic Cameras Compared to Traditional Techniques in the Monitoring of Large Jellyfish

Acoustic cameras have clear advantages over traditional surveys. For example, anchor drift nets are affected by ocean currents, and the net heights decrease in the sea. The rate of net height deformation under the impact of ocean currents is around 50% when nets are fully extended, according to the experience of the local fishers. Thus, the large jellyfish above the anchor drift net in the sea could not be captured. Moreover, the detection blind zone of the acoustic cameras in shallow seas (i.e., a water depth of less than 15 m) is only 1 m at the surface layer, and the detection water layer of the anchor drift net is obviously not as comprehensive as that of an acoustic camera. In this study, compared to the net survey, the acoustic camera enabled more accurate monitoring of the abundance of large jellyfish in the surveyed area, as it can simultaneously scan all of the water layers in shallow waters. Han and Uye [48] found that the abundance of *Aurelia aurita* s.l. measured using an acoustic camera was 3.3 times higher than that measured using plankton nets, thereby offering a more accurate assessment of their abundance. In this study, the average abundance of large jellyfish observed using an acoustic camera was significantly higher than that observed using plankton nets. Similarly, the abundances of *N. nomurai* observed via the acoustic camera in the corresponding water layers in July and August were 8.3 times and 9.2 times higher than those observed using anchor drift nets, respectively. In addition, the abundances of *A. coerulea* were 5.2 times and 5.9 times higher than those observed using anchor drift nets, whereas the abundances of *C. nozakii* were 24.6 times and 6.8 times higher than those observed using anchor drift nets. In this study, the acoustic camera scanning volume in the corresponding water layers (ranging from approximately 8733 to 27,291 m^3 depending on the water depth), over a period of 1 h, was significantly smaller than the scanning volume of the anchor drift net in the 1 h period (approximately 190,800 m^3 in July and 238,500 m^3 in August). Under varying water depth conditions, the volume of seawater filtered by the anchor drift net within the 1 h period was approximately 9–22 times greater than that detected with the acoustic camera in the same period. This may have resulted in more random errors in the acoustic camera monitoring data compared to those derived from the net surveys, particularly when the abundance of large jellyfish was low. We conducted a Pearson correlation analysis between the abundance data of the three types of large jellyfish observed using sonar and the data of those observed with anchor drift nets. The results demonstrated that the abundance data obtained from acoustic camera observations and anchor net surveys for each type of large jellyfish exhibited significant correlations ($p < 0.05$). This confirms that anchor net survey data can serve as reliable baseline data for validating acoustic camera observations of large jellyfish populations. This finding further substantiates the effectiveness of the acoustic camera in accurately monitoring large jellyfish populations. In future studies, extending the observation time and increasing the scanning volume may yield more robust results.

When compared to traditional anchor drift net surveys, there are clear advantages to using acoustic cameras in fish behavioral studies, particularly during nighttime observations or in turbid environments [35,76]. Smith et al. [76] conducted fish surveys on artificial estuarine shorelines in North Carolina, USA, using traditional fish sampling equipment and acoustic cameras. Their preliminary results indicated no significant differences in fish detection numbers between day and night using acoustic cameras, which contrasts with many previous studies that used traditional fish sampling techniques and reported significantly higher catches at night. This suggests that traditional fish sampling methods may underestimate fish numbers during the day because of visual avoidance effects (e.g., foraging, predator avoidance, or breeding). Compared to net and diving surveys, acoustic camera sampling efficiency is less likely to be affected by light availability or stronger daytime fish avoidance behaviors [78], thus resulting in more accurate monitoring of fish behavior. Given the varying water layers in which large jellyfish are distributed, acoustic cameras are clearly a powerful tool for studying the behavior of large jellyfish. In the future, acoustic cameras could be used to observe diurnal movement patterns of large jellyfish in shallow waters.

In terms of the counting method employed here, manual counting was sufficient to meet the data processing requirements because there were relatively few large jellyfish. Similarly to the results of Honda and Watanabe [46], Han and Uye [48], and Makabe et al. [49], we only used manual counting on sonar observation images of large jellyfish and did not employ software-based automatic counting. In cases where a considerable variety and quantity of large jellyfish are observed with an acoustic camera, automatic counting techniques can significantly economize human and material resources. Currently, automatic counting techniques for analyzing acoustic camera observation data obtained through underwater acoustic processing software [79–81], as well as intelligent recognition, classification, and automatic counting techniques developed by convolutional neural networks in machine deep learning [82], are applied in research on fish. Whether these techniques can be applied to the processing of data obtained using acoustic cameras for large jellyfish will be the subject of future studies.

There are certain limitations associated with acoustic cameras with regard to the monitoring of large jellyfish. Acoustic cameras have a detection blind zone of approximately 1 m at the surface, which means that targets located in the surface layer of the water cannot be detected [49,74,83]. As a result, large jellyfish present close to the water's surface are not detected. Conversely, although acoustic cameras also have a detection range limitation (a maximum detection distance of 15 m at a frequency of 1.8 MHz and 35 m at a frequency of 1.1 MHz), this limitation makes them suitable for use in relatively shallow waters. In addition, acoustic cameras have a minimum size limitation. In order to ensure image clarity within the observation range, some applications set a target detection threshold of 4–5 cm [76,83,84]. It is difficult to identify small individuals of large jellyfish species, which could lead to underestimation of their numbers. In addition, anchor drift nets offer several advantages over acoustic cameras. The former method is cost-effective and operationally simple for surveying large jellyfish species. By simultaneously covering a larger volume of seawater over a single period of time, an anchor drift net reduces random errors in the collected data, thereby enhancing the reliability of the survey results. Therefore, the combined use of multiple sampling techniques is effective for fishery resource surveys [76]. Future monitoring of large jellyfish requires a comprehensive approach that uses various techniques to compensate for the shortcomings of each method. Depending on the specific conditions of the survey area, acoustic cameras, traditional nets, and visual surveys can be combined. Acoustic cameras can be used to monitor large jellyfish in complex marine environments or when the number of large jellyfish is too high, causing nets to break and making net surveys challenging.

5. Conclusions

The results of this study demonstrated that acoustic cameras (adaptive resolution imaging sonar) are suitable for monitoring large jellyfish populations in coastal waters. In future work, we will employ an acoustic camera to monitor large jellyfish over a wider range of shallow seas and conduct behavioral studies on them.

Author Contributions: Conceptualization, B.W.; methodology, B.W.; software, X.L.; validation, C.F.; formal analysis, Z.Z.; investigation, B.W., X.L., C.F. and Y.X.; resources, X.L.; data curation, D.Z.;

writing—original draft preparation, B.W.; writing—review and editing, B.W.; visualization, Y.X.; supervision, J.D.; project administration, A.W.; funding acquisition, J.D. All authors have read and agreed to the published version of the manuscript.

Funding: This research was funded by the National Key Research and Development Program of China (No. 2023YFC3108202), and Dalian Science and Technology Innovation Fund project (No. 2022JJ13SN089; No. 2023JJ13SN066).

Institutional Review Board Statement: This study was approved by the former Department of Ocean and Fisheries of Liaoning Province (document number [2018]135).

Informed Consent Statement: Not applicable.

Data Availability Statement: Data are contained within the article.

Acknowledgments: We sincerely appreciate all the reviewers for their invaluable feedback and suggestions.

Conflicts of Interest: The authors declare no conflicts of interest.

References

1. Arai, M.N. Pelagic coelenterates and eutrophication: A review. *Hydrobiologia* **2001**, *451*, 69–87. [CrossRef]
2. Brodeur, R.D.; Sugisaki, H.; Hunt, G.L., Jr. Increases in jellyfish biomass in the Bering Sea: Implications for the ecosystem. *Mar. Ecol. Prog. Ser.* **2002**, *233*, 89–103. [CrossRef]
3. Purcell, J.E. Climate effects on formation of jellyfish and ctenophore blooms. *J. Mar. Biol. Ass. UK* **2005**, *85*, 461–476. [CrossRef]
4. Purcell, J.E.; Uye, S.; Lo, W.T. Anthropogenic causes of jellyfish blooms and their direct consequences for human: A review. *Mar. Ecol. Prog. Ser.* **2007**, *350*, 153–174. [CrossRef]
5. Attrill, M.J.; Wright, J.; Edwards, M. Climate-related increases in jellyfish frequency suggest a more gelatinous future for the North Sea. *Limnol. Oceanogr.* **2007**, *52*, 480–485. [CrossRef]
6. Uye, S. Human forcing of the copepod-fish-jellyfish triangular trophic relationship. *Hydrobiologia* **2011**, *666*, 71–83. [CrossRef]
7. Mills, C.E. Jellyfish blooms: Are populations increasing globally in response to changing ocean conditions? *Hydrobiologia* **2001**, *451*, 55–68. [CrossRef]
8. Kawahara, M.; Uye, S.; Ohtsu, K.; Iizumi, H. Unusual population explosion of the giant jellyfish *Nemopilema nomurai* (Scyphozoa: Rhizostomeae) in East Asian waters. *Mar. Ecol. Prog. Ser.* **2006**, *307*, 161–173. [CrossRef]
9. Uye, S. Blooms of the giant jellyfish *Nemopilema nomurai*: A threat to the fisheries sustainability of the East Asian Marginal Seas. *Plankton Benthos Res.* **2008**, *3*, 125–131. [CrossRef]
10. Richardson, A.J.; Bakun, A.; Hays, G.C.; Gibbons, M.J. The jellyfish joyride: Causes; consequences and management responses to a more gelatinous future. *Trends Ecol. Evol.* **2009**, *24*, 312–322. [CrossRef]
11. Condon, R.H.; Duarte, C.M.; Pitt, K.A.; Robinson, K.L.; Lucas, C.H.; Sutherland, K.R.; Mianzan, H.W.; Bogeberg, M.; Purcell, J.E.; Decker, M.B.; et al. Recurrent jellyfish blooms are a consequence of global oscillations. *Proc. Natl. Acad. Sci. USA* **2013**, *110*, 1000–1005. [CrossRef] [PubMed]
12. Dong, Z.; Liu, D.; Keesing, J.K. Jellyfish blooms in China: Dominant species; causes and consequences. *Mar. Pollut. Bull.* **2010**, *60*, 954–963. [CrossRef] [PubMed]
13. Zhang, F.; Sun, S.; Jin, X.; Li, C. Associations of large jellyfish distributions with temperature and salinity in the Yellow Sea and East China Sea. *Hydrobiologia* **2012**, *690*, 81–96. [CrossRef]
14. Sun, S.; Zhang, F.; Li, C.; Wang, S.; Wang, M.; Tao, Z.; Wang, Y.; Zhang, G.; Sun, X. Breeding places; population dynamics; and distribution of the giant jellyfish *Nemopilema nomurai* (*Scyphozoa: Rhizostomeae*) in the Yellow Sea and the East China Sea. *Hydrobiologia* **2015**, *754*, 59–74. [CrossRef]
15. Dong, J.; Wang, B.; Duan, Y.; Yoon, W.D.; Wang, A.; Liu, X.; Li, Y.; Sun, M.; Chai, Y. Initial occurrence; ontogenic distribution-shifts and advection of *Nemopilema nomurai* (*Scyphozoa: Rhizostomeae*) in Liaodong Bay, China, from 2005–2015. *Mar. Ecol. Prog. Ser.* **2018**, *591*, 185–197. [CrossRef]
16. Graham, W.M.; Martin, D.L.; Martin, J.C. In situ quantification and analysis of large jellyfish using a novel video profiler. *Mar. Ecol. Prog. Ser.* **2003**, *254*, 129–140. [CrossRef]
17. Purcell, J.E. Predation on zooplankton by large jellyfish (*Aurelia labiata; Cyanea capillata; Aequorea aequorea*) in Prince William Sound; Alaska. *Mar. Ecol. Prog. Ser.* **2003**, *246*, 137–152. [CrossRef]
18. Doyle, T.K.; Houghton, J.D.R.; Buckley, S.M.; Hays, G.C.; Davenport, J. The broad-scale distribution of five jellyfish species across a temperate coastal environment. *Hydrobiologia* **2007**, *579*, 29–39. [CrossRef]

19. Bastian, T.; Haberlin, D.; Purcell, J.E.; Hays, G.C.; Davenport, J.; McAllen, R.; Doyle, T.K. Large-scale sampling reveals the spatio-temporal distributions of the jellyfish *A. coerulea* and *Cyanea capillata* in the Irish Sea. *Mar. Biol.* **2011**, *158*, 2639–2652. [CrossRef]
20. Randriarilala, F.; Kitakado, T.; Shiode, D.; Sakaguchi, M.; Hayashi, T.; Tokai, T. Density estimation of the giant jellyfish *Nemopilema nomurai* around Japan using an alternative modified detection function for left truncation in a line transect survey. *Fish. Sci.* **2014**, *80*, 261–271. [CrossRef]
21. Yoon, W.D.; Hahn, K.; Dong, J.; Chae, J. Monthly geographical distribution of *Nemopilema nomurai* (*Scyphozoa rhizosotomeae*) in the Bohai and North Yellow Seas: A preliminary study using ships of opportunity. *Ocean Sci. J.* **2018**, *53*, 699–706. [CrossRef]
22. Houghton, J.D.R.; Doyle, T.K.; Davenport, J.; Hays, G.C. Developing a simple; rapid method for identifying and monitoring jellyfish aggregations from the air. *Mar. Ecol. Prog. Ser.* **2006**, *314*, 159–170. [CrossRef]
23. Yang, D.; Cui, W.; Zhang, H.; Xu, Z.; Zhang, A. Application of new technology in jellyfish monitoring. *Ocean Dev. Manag.* **2014**, *4*, 38–41.
24. Hirose, M.; Mukai, T.; Hwang, D.; Lida, K. The acoustic characteristics of three jellyfish species: *Nemopilema nomurai; Cyanea nozakii* and *A. coerulea*. *ICES J. Mar. Sci.* **2009**, *66*, 1233–1237. [CrossRef]
25. Brierley, A.S.; Boyer, D.S.; Axelsen, B.E.; Lynam, C.P.; Sparks, C.A.J.; Boyer, H.J.; Gibbons, M.J. Towards the acoustic estimation of jellyfish abundance. *Mar. Ecol. Prog. Ser.* **2005**, *295*, 105–111. [CrossRef]
26. Colombo, G.A.; Benović, A.; Malej, A.; Lučić, D.; Makovec, T.; Onofri, V.; Acha, M.; Madirolas, A.; Mianzan, H. Acoustic survey of a jellyfish-dominated ecosystem (Mljet Island; Croatia). *Hydrobiologia* **2009**, *616*, 99–111. [CrossRef]
27. Gorbatenko, K.M.; Nikolayev, A.V.; Figurkin, A.L.; Il'inskii, E.N. Quantitative composition; distribution and feeding of large jellyfish (Scyphozoa et Hydrozoa) on the west Kamchatka shelf in summer. *Russ. J. Mar. Biol.* **2009**, *35*, 579–592. [CrossRef]
28. Graham, T.R.; Harvey, J.T.; Benson, S.R.; Renfree, J.S.; Demer, D.A. The acoustic identification and enumeration of scyphozoan jellyfish; prey for leatherback sea turtles (*Dermochelys coriacea*); off central California. *ICES J. Mar. Sci.* **2010**, *67*, 1739–1748. [CrossRef]
29. Robertis, A.D.; Taylor, K. In situ target strength measurements of the scyphomedusa *Chrysaora melanaster*. *Fish. Res.* **2014**, *153*, 18–23. [CrossRef]
30. Moursund, R.A.; Carlson, T.J.; Peters, R.D. A fisheries application of a dual-frequency identification sonar acoustic camera. *ICES J. Mar. Sci.* **2003**, *60*, 678–683. [CrossRef]
31. Tiffan, K.F.; Rondorf, D.W.; Skalicky, J.J. Imaging fall Chinook salmon redds in the Columbia River with a dual-frequency identification sonar. *N. Am. J. Fish. Manag.* **2004**, *24*, 1421–1426. [CrossRef]
32. Baumgartner, L.J.; Reynoldson, N.; Cameron, L.; Stanger, J. *Assessment of a Dual-Frequency Identification Sonar (DIDSON) for Application in Fish Migration Studies*; Fisheries Final Report Series 84; NSW Department of Primary Industries: Narrandera, NSW, Australia, 2006.
33. Becker, A.; Cowley, P.D.; Whitfield, A.K.; Järnegren, J.; Næsje, T.F. Diel fish movements in the littoral zone of a temporarily closed South African estuary. *J. Exp. Mar. Biol. Ecol.* **2011**, *406*, 63–70. [CrossRef]
34. Crossman, B.J.A.; Martel, G.; Johnson, P.N.; Bray, K. The use of Dual-frequency IDentification SONar (DIDSON) to document white sturgeon activity in the Columbia River; Canada. *J. Appl. Ichthyol.* **2011**, *27* (Suppl. S2), 53–57. [CrossRef]
35. Lee, D.; Oh, W.; Gim, B.M.; Lee, J.S.; Yoon, E.; Lee, K. Investigating the effects of different led wavelengths on aggregation and swimming behavior of chub mackerel (*Scomber japonicus*). *Ocean Sci. J.* **2019**, *54*, 573–579. [CrossRef]
36. Holmes, J.A.; Cronkite, G.M.; Enzenhofer, H.J.; Mulligan, T.J. Accuracy and precision of fish-count data from a "dual-frequency identification sonar" (DIDSON) imaging system. *ICES J. Mar. Sci.* **2006**, *63*, 543–555. [CrossRef]
37. Maxwell, S.L.; Gove, N.E. Assessing a dual frequency identification sonar's fish-counting accuracy, precision, and turbid river range capability. *J. Acoust. Soc. Am.* **2007**, *122*, 3364–3377. [CrossRef]
38. Mueller, A.; Mulligan, T.J.; Withler, P.K. Classifying sonar images: Can a computer-driven process identify eels? *N. Am. J. Fish. Manag.* **2008**, *28*, 1876–1886. [CrossRef]
39. Magowan, K.; Reitsma, J.; Murphy, D. Use of Dual-Frequency Identification Sonar to monitor adult river herring in a small coastal stream. *Mar. Coast. Fish. Dyn. Manag. Ecosyst. Sci.* **2012**, *4*, 651–659. [CrossRef]
40. Petreman, I.C.; Jones, N.E.; Milne, S.W. Observer bias and subsampling efficiencies for estimating the number of migrating fish in rivers using Dual-frequency IDentification SONar (DIDSON). *Fish. Res.* **2014**, *155*, 160–167. [CrossRef]
41. Hughes, J.B.; Hightower, J.E. Combining split-beam and dual-frequency identification sonars to estimate abundance of anadromous fishes in the Roanoke river, North Carolina. *N. Am. J. Fish. Manag.* **2015**, *35*, 229–240. [CrossRef]
42. Burwen, D.L.; Fleischman, S.J.; Miller, J.D. Accuracy and Precision of Salmon Length Estimates Taken from DIDSON Sonar Images. *Trans. Am. Fish. Soc.* **2010**, *139*, 1306–1314. [CrossRef]
43. Xu, C.H.; Mizuno, K.; Asada, A.; Abukawa, K.; Yamamuro, M. 3D views generation and species classification methods of aquatic plants using acoustic images. *J. Mar. Acoust. Soc. Jpn.* **2013**, *40*, 14–26. [CrossRef]

44. Mizuno, K.; Abukawa, K.; Kashima, T.; Asada, A.; Fujimoto, Y.; Shimada, T. Quantification of whooper swan damage to lotus habitats using high-resolution acoustic imaging sonar in Lake Izunuma; Japan. *Aquat. Bot.* **2013**, *110*, 48–54. [CrossRef]
45. Mizuno, K.; Abukawa, K.; Kashima, T.; Asada, A.; Fujimoto, Y.; Shimada, T. Assessing the biological process of *Hydrilla verticillata* predation in a eutrophic pond using high-resolution acoustic imaging sonar. *Limnology* **2016**, *17*, 13–21. [CrossRef]
46. Honda, N.; Watanabe, T. Observation of the giant jellyfish *Nemopilema nomurai* using an underwater acoustic camera. *Nippon Suisan Gakkaishi* **2007**, *73*, 919–921. [CrossRef]
47. Lee, K.; Bae, B.S.; Kim, I.O.; Yoon, W.D. Measurement of swimming speed of giant jellyfish *Nemopilema nomurai* using acoustics and visualization analysis. *Fish Sci.* **2010**, *76*, 893–899. [CrossRef]
48. Han, C.H.; Uye, S. Quantification of the abundance and distribution of the common jellyfish *Aurelia aurita* s.l. with a Dual-frequency IDentification SONar (DIDSON). *J. Plankton Res.* **2009**, *31*, 805–814. [CrossRef]
49. Makabe, R.; Kurihara, T.; Uye, S. Spatio-temporal distribution and seasonal population dynamics of the jellyfish *Aurelia aurita* s.l. studied with Dual-frequency IDentification SONar (DIDSON). *J. Plankton Res.* **2012**, *34*, 936–950. [CrossRef]
50. Tong, J.; Han, J.; Shen, W. Preliminary research on the image processing of acoustic camera and its application in fishery. *Hunan Agric. Sci.* **2010**, *17*, 149–152; 156.
51. Tong, J.; Han, J.; Asada, A.; Masahiko, M. Counting method of upstream juvenile ayu (*Plecoglossus altivelis*) by acoustic camera. *Fish. Mod.* **2009**, *36*, 29–33.
52. Zhang, J. *Researches on DIDSON-Based Fish Quantitative Assessment Technology*; Shanghai Ocean University: Shanghai, China, 2012.
53. Zhang, H.; Wei, Q.; Kang, M. Measurement of swimming pattern and body length of cultured Chinese sturgeon by use of imaging sonar. *Aquaculture* **2014**, *434*, 184–187. [CrossRef]
54. Shen, W.; Peng, Z.; Gong, X.; Zhang, J.; Zhu, Z. Detection of fish resources in reservoir and spatial and temporal analysis based on DIDSON. *J. Shanghai Ocean Univ.* **2022**, *31*, 1457–1466.
55. Zhang, P.; Qiao, Y.; Jin, Y.; Lek, S.; Yan, T.; He, Z.; Chang, J.; Cai, L. Upstream migration of fishes downstream of an under-construction hydroelectric dam and implications for the operation of fish passage facilities. *Glob. Ecol. Conserv.* **2020**, *23*, e01143. [CrossRef]
56. Cheng, J.; Ding, F.; Li, S.; Yan, L.; Ling, J.; Li, J.; Liu, Y. A study on the quantity distribution of macro-jellyfish and its relationship to sea water temperature and salinity in the East China Sea Region. *Acta Ecol. Sin.* **2005**, *25*, 440–445.
57. Wang, P.; Zhang, F.; Sun, S.; Yang, T. Distribution of gaint jellyfish in the Bohai Sea in June 2018. *Oceanol. Limnol. Sin.* **2018**, *51*, 85–94.
58. Jia, C.; Fujimori, Y.; Wang, X.; Guan, C. Performance of large-scale stow nets for investigating jellyfsh. *Fish Sci.* **2023**, *89*, 595–603. [CrossRef]
59. Wang, S.; Zhang, G.; Sun, S.; Wang, Y.; Zhao, Z. Population dynamics of three Scyphozoan Jellyfish species during summer of 2011 in Jiaozhou Bay. *Oceanol. Limnol. Sin.* **2012**, *43*, 471–479.
60. Qiu, Y. *Application of UAV Low-Altitude Remote Sensing in Monitoring Large Jellyfish and Beach Litter*; University of Chinese Academy of Sciences: Qingdao, China, 2022.
61. Zhang, Y.; Li, G.; Wang, Z.; Jia, B.; Wang, Z.; Yong, J.; Wang, R.; Jiang, Y. A method of jellyfish detection based on acoustic image. *Tech. Acoust.* **2016**, *35*, 79–80.
62. Wang, B.; Fang, L.; Dong, J.; Li, Y.; Liu, X.; Li, Y.; Sun, M.; Wang, W. Review of acoustic techniques in the monitoring and assessment of the giant jellyfish. *Acta Ecol. Sin.* **2017**, *37*, 8187–8196.
63. Wang, Z.; Fu, Y. Monitoring of cold source biomass at water intake of Hongyanhe Nuclear Power Plant by using acoustic methods. *Sci. Technol. Innov.* **2021**, *36*, 169–172.
64. Fu, Y.; Tang, Y.; Wang, S.; Meng, W.; Wang, S.; Liu, X. Flux measurement of jellyfish near Hongyanhe Nuclear Power Station area in summer by using acoustic methods. *J. Dalian Ocean Univ.* **2021**, *36*, 325–333.
65. Dong, J.; Jiang, L.; Sun, M.; Wang, B.; Li, Y.; Tan, K.; Sun, S.; Chai, Y. *Biology of Large Jellyfish in the Bohai Sea and Northern Yellow Sea*; Ocean Press: Beijing, China, 2013; pp. 1–265.
66. Ge, L.J.; He, D.M. A report on *Cyanea* sp. blooming and the catch of edible jellyfish *Rhopilema esculentum* being reduced suddenly. *Chin. Fish.* **2005**, *9*, 23–24.
67. Wang, B.; Dong, J.; Liu, C.; Sun, M.; Yu, X.; Liu, X.; Li, Y.; Li, P. Distribution of giant jellyfish and major zooplankton in jellyfish release area of Liaodong Bay, Bohai Sea in early summer. *Prog. Fish. Sci.* **2010**, *31*, 83–90.
68. Wang, B.; Dong, J.; Wang, W.; Li, Y.L.; Li, Y.; Liu, X.; Fu, J. The quantity distribution of giant jellyfish and its relationship to seawater temperature and salinity in inshore waters of the Northern Liaodong Bay Region. *Oceanol. Limnol. Sin.* **2012**, *43*, 568–578.
69. Wang, B.; Qin, Y.B.; Dong, J.; Li, Y.L.; Wang, W.; Li, Y.; Sun, M.; Liu, C. Dynamic distribution of *Nemopilema nomurai* in inshore waters of the northern Liaodong Bay, Bohai Sea. *Acta Ecol. Sin.* **2013**, *33*, 1701–1712. [CrossRef]
70. Wang, B.; Li, Y.; Shen, H.; Li, Y.; Wang, W.; Sun, M.; Dong, J. Quantity distribution of *Cyanea nokazii* in inshore waters of northern Liaodong Bay, Bohai Sea in 2005–2013. *Mar. Fish.* **2014**, *36*, 146–154.

71. Wang, B.; Liu, X.; Wang, X.; Ji, G.; Dong, J. Application of Dual-Frequency Identification Sonar (DIDSON) to monitor giant jellyfish *Nemopilema nomurai* in inshore waters of Liaodong Bay. *J. Fish. Sci. China* **2022**, *29*, 1223–1235.
72. *Local Standards of Liaoning Province DB21/T 2823-2017*; Technical Specification for *Nemopilema nomurai* Monitoring Survey. Liaoning Provincial Bureau of Quality and Technical Supervision: Shenyang, China, 2017; pp. 1–9.
73. *Local Standards of Liaoning Province DB21/T 3368-2021*; Technical Specification for Monitoring Survey and Biomass Assessment of Disaster-Causing Jellyfish in Offshore Areas. Liaoning Provincial Market Supervision and Administration Bureau: Shenyang, China, 2021; pp. 1–6. (In Chinese)
74. Mo, W.; Wang, C.; Qin, X.; Zhang, M.; Liu, H. Acoustic Monitoring on Fish Resources in the Dongqing and Guangzhao Reservoirs of Beipan River. *J. Hydroecology* **2015**, *36*, 10–17. (In Chinese)
75. Jůza, T.; Rakowitz, G.; Draštík, V.; Blabolil, P.; Herzig, A.; Kratochvíl, M.; Muška, M.; Říha, M.; Sajdlová, Z.; Kubečka, J. Avoidance reactions of fish in the trawl mouth opening in a shallow and turbid lake at night. *Fish. Res.* **2013**, *147*, 154–160. [CrossRef]
76. Smith, C.S.; Paxton, A.B.; Donaher, S.E.; Kochan, D.P.; Neylan, I.P.; Pfeifer, T.; Hoeck, R.V.V.; Taylor, J.C. Acoustic camera and net surveys reveal that nursery enhancement at living shorelines may be restricted to the marsh platform. *Ecol. Eng.* **2021**, *166*, 106232. [CrossRef]
77. Mutlu, E. Target strength of the common jellyfish (*Aurelia aurita*): A preliminary experimental study with a dual-beam acoustic system. *ICES J. Mar. Sci.* **1996**, *53*, 309–311. [CrossRef]
78. Rakowitz, G.; Tušer, M.; Říha, M.; Jůza, T.; Balk, H.; Kubečka, J. Use of high-frequency imaging sonar (DIDSON) to observe fish behaviour towards a surface trawl. *Fish. Res.* **2012**, *123–124*, 37–48. [CrossRef]
79. Cook, D.; Middlemiss, K.; Jaksons, P.; Davison, B.; Jerrett, A. Validation of fish length estimations from a high frequency multi-beam sonar (ARIS) and its utilisation as a field-based measurement technique. *Fish. Res.* **2019**, *218*, 59–68. [CrossRef]
80. Han, J.; Honda, N.; Asada, A.; Shibata, K. Automated acoustic method for counting and sizing farmed fish during transfer using DIDSON. *Fish. Sci.* **2009**, *75*, 1359–1367. [CrossRef]
81. Kang, M. Semiautomated Analysis of Data from an Imaging Sonar for Fish Counting, Sizing, and Tracking in a Post-Processing Application. *Fish. Aquat. Sci.* **2011**, *14*, 218–225. [CrossRef]
82. Connolly, R.M.; Jinks, K.I.; Shand, A.; Taylor, M.D.; Gaston, T.F.; Becker, A.; Jinks, E.L. Out of the shadows: Automatic fish detection from acoustic cameras. *Aquat. Ecol.* **2023**, *57*, 833–844. [CrossRef]
83. Zhou, J.; Wang, C.; Liu, D.; Huang, Y.; Xiang, J. Hydro-acoustic investigation and assessment on fishes in the near downstream of Gezhouba Dam. *Resour. Environ. Yangtze Basin* **2014**, *23*, 1551–1557.
84. Shen, W.; Zhu, Z.; Zhang, J.; Cao, Z.; Peng, Z. Fish target recognition and counting based on Dual-frequency Identification Sonar. *Fish. Mod.* **2020**, *47*, 81–86.

MDPI AG
Grosspeteranlage 5
4052 Basel
Switzerland
Tel.: +41 61 683 77 34

Fishes Editorial Office
E-mail: fishes@mdpi.com
www.mdpi.com/journal/fishes

www.ingramcontent.com/pod-product-compliance
Lightning Source LLC
Chambersburg PA
CBHW061622210926
51850CB00009B/2403